Damage to Concrete Structures

Damage to Concrete Structures

Geert De Schutter

CRC Press
Taylor & Francis Group
Boca Raton London New York

CRC Press is an imprint of the
Taylor & Francis Group, an **informa** business

A SPON PRESS BOOK

CRC Press
Taylor & Francis Group
6000 Broken Sound Parkway NW, Suite 300
Boca Raton, FL 33487-2742

Printed in the United States of America on acid-free paper
Version Date: 20120523

International Standard Book Number: 978-0-415-60388-1 (Paperback)

Library of Congress Cataloging-in-Publication Data

De Schutter, Geert.
    Damage to concrete structures / Geert De Schutter.
        p. cm.
    Summary: "Unlike the more specialist books which deal with diagnosis techniques or non-destructive testing, with repair and strengthening of concrete structures or with specific degradation mechanisms such as sulfate attack or alkali-silica reaction, this book describes and explains the different types of damage to concrete structures comprehensively. It is written as a textbook for undergraduate and masters students, and is also helpful to practitioners such as design engineers, architects or consultants"-- Provided by publisher.
    Includes bibliographical references and index.
    ISBN 978-0-415-60388-1 (pbk.)
    1. Concrete construction. 2. Building failures--Prevention. 3. Concrete--Testing. I. Title.

TA681.D397 2012
624.1'834--dc23

2012017947

Visit the Taylor & Francis Web site at
http://www.taylorandfrancis.com

and the CRC Press Web site at
http://www.crcpress.com

To my father-in-law,

who passed away, too young.

Human life is more complex than concrete.

Contents

Foreword

Not only is *Damage to Concrete Structures* up-to-date with its coverage of the topic, it is also unique in that it addresses a wide range of subjects that influence the performance of concrete structures in service. As a result, it encompasses many aspects of concrete construction beyond the typical materials and structural design courses taught in universities, including informative material that is rarely detailed in textbooks—especially the chapters on Inappropriate Design and Errors during Casting. It should therefore be a valuable resource for students as they need to be knowledge-able in all the important issues that can affect the successful achievement of concrete structures that will be durable and, therefore, more sustainable.

Professor R. Douglas Hooton
NSERC/CAC Industrial Research
Chair in Concrete Durability and Sustainability
University of Toronto
April 2012

Endorsement

This text book is endorsed by RILEM, after review by Prof. R. Doug Hooton, on behalf of RILEM Educational Activities Committee (EAC). RILEM, The International Union of Laboratories in Construction Materials, Systems and Structures, founded in 1947, is a non-governmental technical association whose vocation is to contribute to progress in the construction sciences, techniques and industries, essentially by means of the communication it fosters between research and practice. RILEM's focus is on construction materials and their use in building and civil engineering structures, covering all phases of the building process from manufacture to use and recycling of materials. More information on RILEM and its publications can be found on www.RILEM.net.

Preface

The human perception of concrete is not always founded on clear and accurate arguments. On the one hand, concrete structures tend to be perceived as gray and dull with a high carbon footprint. This perception is mainly based on historical reasons, and does not duly consider recent evolutions in concrete technology, which enable the completion of attractive and nicely shaped, durable, and sustainable concrete structures. On the other hand, concrete is perceived as being a strong, solid, and protective material. When something is 'cast in concrete,' it will be there forever. However, although we surely can make durable and long-lasting concrete structures, they are not built for eternity. Things can go wrong at different stages of the construction process or during service.

This book explains how damage can occur to concrete structures. The contents of the different chapters are structured along the timing of the actions leading to damage. Damage can occur due to inappropriate design, errors during execution, some mechanisms occurring during hardening of the concrete, or actions or degradation mechanisms during service life (hardened concrete). The different actions or mechanisms are explained in a fundamental way without too many physical or chemical details. The degradation mechanisms are illustrated with many drawings and photographs taken of real structures.

In the first instance, the book aims at students (bachelor or master level) and gives a general introduction to concrete damage and durability. It is written in such a way that it can be used in educational programs at universities or colleges in the fields of civil engineering, structural engineering, architectural engineering, and environmental engineering. Additionally, the book will be very helpful for professionals including architects, design engineers, experts and consultants. These readers will find this book provides a helpful overview of the different damage mechanisms.

Although this book presents a general introduction, it can also benefit doctoral and post-doctoral researchers as a resource referencing other, more detailed scientific publications that offer more advanced information on some specific degradation mechanisms.

Upon completion of this textbook, I am very grateful to many of my (former) doctoral students, and to numerous colleagues worldwide from whom I have learned a lot. In this textbook, I have tried to summarize the state-of-the-art information on the degradation of concrete structures, and also to give a clear and comprehensive overview of what can go wrong. By the specialists in the various sub-domains, the information will most probably be considered as a somewhat simplified and condensed version of the actual scientific knowledge. Nevertheless, I am confident that within this process of summarizing existing knowledge and presenting it to bachelor and master level students, no confusion has been introduced. On the contrary, this textbook explains in a clear and pedagogical way the basic knowledge on damage to concrete structures.

To all those who contributed and are still contributing to the advancement of knowledge in this domain, I am very grateful. It was my honour and privilege to summarize a substantial part of this knowledge, hoping to motivate bachelor and master students to learn from it, and possibly to encourage them to contribute themselves to future developments, and to durable and sustainable concrete structures.

Geert De Schutter
Ghent, January 2012

About the Author

Geert De Schutter is a full professor at Ghent University, Belgium. He is currently conducting research in the field of concrete technology at the Magnel Laboratory for Concrete Research, Department of Structural Engineering. He is laureate of several national and international awards including the Vreedenburgh Award in 1998 and the prestigious international RILEM Robert L'Hermite Medal in 2001. In 2002, he was an invited professor at Oita University, Japan. From 2008 to 2011, he was an invited professor at the University of Cergy-Pontoise near Paris, France. In 2012, he was awarded the Francqui Chair at the University of Liège, Belgium. Since 1 February 2009, Professor De Schutter has served as director of development of RILEM. He previously co-authored the textbook *Self-Compacting Concrete* published by Whittles Publishing in 2008.

Chapter 1

Durability and service life

1.1 INTRODUCTION

For many years after the introduction of reinforced concrete as a construction material at the end of the nineteenth century, it was considered that concrete structures were built to last without any further maintenance or repair. Since then, however, we have learned that this unfortunately is not the case. Serious degradation mechanisms can severely reduce the service life of concrete structures: steel reinforcement can corrode, cement matrix can be attacked, and even aggregates can show detrimental processes. In many cases, the remaining service life of damaged structures can be upgraded by a combined action of repair and maintenance. However, in some cases demolition and replacement are the only options.

The question might be raised why ancient cultures were apparently able to build structures which are still in good shape after several thousands of years, while modern structures can show degradation after only a few decades or even earlier. A marvellous example of a structure that seems to last forever is the Roman aqueduct Pont du Gard (Figure 1.1) in the South of France, constructed about 2000 years ago by means of blocks of limestone. Although the structure still looks marvellous and is impressive, it is clear that even this cultural heritage is suffering from degradation mechanisms that have attacked the limestone (Figure 1.2). Furthermore, it should also be mentioned that we all know the Pont du Gard only because it has survived until now. Many Roman aqueducts did not survive and are long forgotten or known to very few thanks to some remaining ruins, as nicely illustrated by the Roman aqueduct near Fontvieille (Figure 1.3), not far from the Pont du Gard.

Nevertheless, it is clear that ancient cultures sometimes built structures capable of incredible performance when considering their heritage. Modern structures seem to be more prone to degradation. Several reasons can be listed to explain why durability issues related to concrete structures have become more prominent during the last decades:

Figure 1.1 Pont du Gard, France, general view of the Roman aqueduct.

Figure 1.2 Pont du Gard, France, degradation of the limestone.

Figure 1.3 Some leftovers of a Roman aqueduct near Fontvieille, France.

- Reinforced concrete is a relatively recent construction material. Although first patents date from the middle of the nineteenth century (patent of J. Lambot in 1855 and of J. Monier in 1867), the application in buildings and structures started to grow only after 1900. It became a mainstream construction material only after World War II.
- After World War II, the construction industry in Europe showed an exponential growth in the 1960s and 1970s. The reconstruction of Europe was of major importance, and the main focus was on a timely completion of buildings and structures rather than on quality and durability, making them more vulnerable to degradation. Most of these structures have now reached a service life of fifty years.
- Many degradation mechanisms require a time period of about twenty years before damage becomes visible. It is not a coincidence that some physical or chemical degradation mechanisms have only been recognized twenty years after a strong growth of the construction activity. One example is alkali silica reaction (ASR). In Belgium, as in many industrialized countries, ASR was first diagnosed in the 1980s. Only twenty years after the construction boom of the 1960s it became clear that many concrete structures suffered from ASR. As we now better understand the mechanisms leading to ASR damage, measures have been defined to avoid ASR damage as much as possible in new concrete structures.
- Due to increased industrial activity, the aggressiveness of the environment has increased. Waterways and atmosphere have become more

and more polluted, giving rise to an increased degradation risk to (concrete) structures.

- More advanced design methods, e.g. ultimate limit state design methods instead of elastic design, and advanced developments on the materials level, such as the implementation of modern plasticizers, have led to more slender concrete structures. As a consequence, the structures might be somewhat more sensitive to degradation mechanisms compared to over-designed structures.

Nowadays, the attention given to durability (and sustainability) is at a much higher level than a few decades ago. This textbook aims to further transfer the knowledge on the possible degradation mechanisms of concrete structures. However, first some general information is given on durability and service life of structures in general.

1.2 DURABILITY AND SERVICE LIFE

Structures in general, and concrete structures in particular are designed to fulfil requirements related to strength and function during a certain time period, without incurring unexpected costs for maintenance or repair. This time period is called the service life of the structure, and it is mainly determined by the structure's durability. The durability is the property of the material or the structure to withstand serious degradation mechanisms, making sure that the decline of the initial properties is kept within acceptable quality limits. In other words, the durability of a structure is the capacity to resist degradation, making sure that the required service life can be reached.

After successful completion of a structure, the initial quality is higher than the minimal required level as shown in Figure 1.4. With time, due to the 'ageing' of the structure, the quality of the structure decreases. In normal conditions, the decreasing quality reaches the minimal required level only at the end of the target service life. In the case of aggressive environment, the rate of quality loss can be very high, possibly leading to an accelerated crossing of the minimal required level, meaning that the target service life is not reached. In this case, repair actions can be necessary to reach the target service life. The effect of repair is visible in Figure 1.4 by a sudden increase in quality of the structure. Distinction can be made between corrective repair in order to rehabilitate the structure after an unforeseen accelerated degradation, or preventive repair anticipating predicted future damage. In principle, for every important structure a specific maintenance and repair strategy should be defined and linked with an adequate inspection and monitoring strategy.

Reaching the end of service life does not necessarily mean that the structure will collapse or will have to be demolished. It rather means that

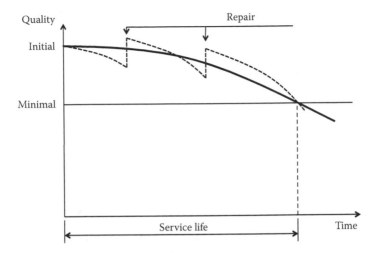

Figure 1.4 Schematic representation of the service life and the effect of repair.

the safety of the structure might no longer be at a very high level, or that high maintenance or repair costs will have to be expended in order to keep the performance of the structure at an acceptable functional level. Quite often it is a matter of economical and profit calculations in order to decide whether to maintain and repair or to demolish and reconstruct. Another option is to give a new destination to the structure, with possibly a reduced required minimal quality level.

The service life of a structure can be studied from at least three different points of view: technical or structural, functional, and economical. Depending on the point of view, a technical, functional, or economical service life can be defined. Technical or structural requirements are linked with the strength of materials and the load bearing capacity of structures. These requirements are commonly considered in design standards and model codes.

Functional requirements are related to the normal use of a building or a structure. The width of a bridge and the height underneath must be in accordance with the dimensions of the vehicles crossing the bridge or driving under the bridge. The functional service life is not determined by load bearing capacity or other technical issues, but rather by the evolution in traffic. As an example, the Walnut Lane Bridge (which was the first prestressed concrete bridge in the United States of America, designed by G. Magnel, and built in 1949 in Philadelphia) was replaced by a larger bridge in the 1990s. At that time, the Walnut Lane Bridge was still in very good shape and showed no problems with load bearing capacity. It was rather the intensified traffic, needing more lanes, which caused the replacement of this famous bridge. Other examples are the many office buildings in cities like Brussels, which are completely renovated after only fifteen to twenty years of service because of the evolution in functional office requirements (e.g. new communication methods, landscape offices instead of separate offices).

From the economical point of view, a structure can be considered as an investment. In this case, the requirements are defined on the levels of 'profit' and 'margins'. One example is the ferry boats, once linking mainland Europe with the British Island. After the completion of the Channel tunnel, many ferry boats were taken out of the fleet. Although these boats still perfectly complied with technical or functional requirements, from an economical point of view they became obsolete as a substantial part of traffic shifted towards the tunnel.

The further detailed treatment in this textbook of the degradation mechanisms in concrete structures is more linked to technical or structural requirements. Many factors can influence the structural response of a structure as illustrated in Figure 1.5. First, based on structural requirements, the design stage is very important. During design, the overall dimensions are defined as well as some important construction details. The appropriate materials have to be selected. In the case of concrete structures, this means an appropriate mix design, based on carefully selected constituent materials. All of this should be in good agreement with the future environment of the structure to be designed.

In the construction stage, when the structure is actually built, some deviations can occur in comparison with the design; e.g. geometry can be

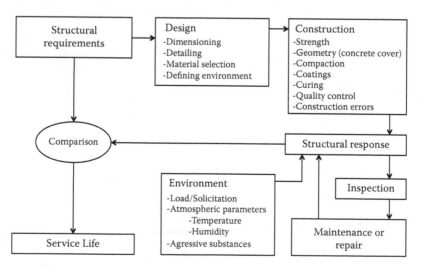

Figure 1.5 Structural response and service life.

slightly different and a real strength is reached instead of a design strength. For the specific case of concrete structures, the concrete cover, the compaction of the concrete, and the curing of the concrete are of major importance for the final durability characteristics of the completed structure. When coatings have to be applied, the real performance of this coating is also very important, as it typically has to provide some protection. The quality control on the construction site and the actions to avoid construction errors are important issues during the construction stage.

After completion of the structure, a certain structural response will be reached. This response could be experimentally verified, e.g. trial loading of a bridge. When design and construction have been properly done, the initial structural response should be better than the structural requirements on which the design was based. From that point on, the environment starts to have an influence on the structure. Due to mechanical, physical, and chemical loading, the structure will start to degrade. Within this process, the atmospheric conditions (temperature and humidity) and the presence of potentially aggressive substances largely influence the degradation rate.

Regular inspection or continuous monitoring will reveal relevant information on the evolution of the structural response of the ageing structure. Depending on the outcome of the comparison with the structural requirements, maintenance or repair actions could be defined in order to reach the target service life. When the structural response is no longer in agreement with the structural requirements, the end of service life is reached.

1.3 STRATEGIES AND COST

Although it is the intention to further focus on technical requirements, it is certainly good to know that technical choices have economical consequences. This will be illustrated in this section by considering the example of reinforcement corrosion in concrete structures. While the scientific background concerning reinforcement corrosion will be explained in Chapter 5, we illustrate some cost strategies for the case of reinforcement corrosion by considering the generally accepted Tuutti model as shown in Figure 1.6. The vertical axis shows the evolution of corrosion, considering an induction period (horizontal line) and a propagation period (inclined line). After a certain time period, the corrosion level reaches a critical corrosion level, defining the end of service life.

Four different strategies can be followed in order to reach a target service life.

Strategy A is the strategy of good practice, involving an adequate design and high quality execution. Within this strategy, an appropriate durability

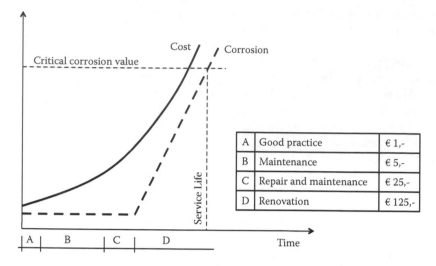

Figure 1.6 Cost strategies.

A	Good practice	€ 1,-
B	Maintenance	€ 5,-
C	Repair and maintenance	€ 25,-
D	Renovation	€ 125,-

level of the structure is reached by carefully considering all influencing factors as previously described.

Strategy B relies on maintenance. It is accepted that the initiation of corrosion progresses somewhat further by penetration of carbon dioxide into the concrete cover. Once a certain carbonation depth is reached, some maintenance will be done by the application of coatings, or by increasing the concrete cover thickness by applying an extra mortar layer.

Strategy C accepts that the carbonation depth reaches the reinforcement, so that the corrosion process actually can start degrading the reinforcement. As a consequence of the expansive formation of corrosion products, cracks can be initiated in the concrete cover. Within this strategy, repair activities will be needed, possibly combined with further maintenance.

Following strategy D, no action is taken until severe damage has occurred to the concrete structure, including spalling of concrete cover due to expansive action of the corrosion products. The reinforcement bars are severely corroded, over larger areas. At this stage, important and expensive renovation and rehabilitation actions are needed, involving partial or full replacement of structural elements.

The costs involved with the different strategies are shown in a relative way in Figure 1.6. Shifting the strategy one level implies a multiplication of the total costs by a factor of five. In this way, it is illustrated that the same service life can be obtained by a relatively small initial investment following the good practice, or by a relatively high investment by relying on repair and rehabilitation.

In case repair actions are needed, it should be realized that a qualitative and durable repair is based on a combination of the following important elements:

- The repair materials should comply with some minimal performance criteria, and preferably undergo some production and compliance testing. Not only the individual materials have to be tested, but also the entire repair system obtained by the consecutive application of different materials in different layers, within so-called repair systems.
- The technicians applying the repair materials should be well-trained and have sufficient knowledge of concrete in general and damage to concrete more particularly.
- For important structures to be repaired, an independent control should be performed, not only concerning the repair itself, but also concerning the diagnosis to be performed before defining the adequate repair method.

1.4 DURABLE CONCRETE

The best and most economical option to reach a target service life is to carefully design the structure duly considering the effect of relevant aggressive actions. An important element within this approach of 'good practice' is to design a durable concrete composition. A durable concrete structure starts with a durable concrete for the considered application. Some traditional parameters are generally considered to be important for the durability of concrete: the water/cement ratio, the cement content, and the cement type. Concrete strength can give a good indication, but is certainly not a satisfying condition with regard to durability. Often neglected is the influence of the maximum particle size on the required cement content within a mix. The application of puzzolan and inert fillers, and the degree to which these fillers can be considered as cement replacing materials, is an important point of discussion in concrete practice. The issue of designing a durable concrete mix is further treated in this section. It is also explained that casting and curing operations are also important with respect to the final durability properties of the completed structure. As a general point of attention, it should be realized that the concrete cover is crucial to some durability properties of concrete structures.

1.4.1 Fundamental background of concrete durability

A fundamental study of the durability of concrete has to start from the physical and chemical properties of the material and needs to consider the transport properties of potentially aggressive liquids and gasses in the pore system. A detailed treatment of this fundamental information is beyond the scope of this textbook. Here, the fundamental background will be translated to practical measures which can improve the concrete durability. Nevertheless, in order to fully understand the durability behaviour of concrete, it should be realized that macroscopic or so-called engineering properties of concrete are to a large extent determined by the microstructure of the material.

The microstructure of cementitious materials is generally the result of the chemical reaction between cement and water, called the hydration reaction. Figure 1.7 schematically presents the most important chemical processes occurring during the hydration of Portland cement. It should be mentioned that shorthand cement chemistry notations (Hewlett 1988) are used in this figure, instead of classical chemical notations ($C = CO_2$, $S = SiO_2$, $A = Al_2O_3$, $F = Fe_2O_3$, $\underline{S} = SO_3$).

The durability of concrete is largely determined by the microstructure, not only because of the transport mechanisms within the pore system, but also because of the amount and chemical nature of the formed hydration products. As an example, portlandite plays a key role within the carbonation process, while monosulphate is crucial within the process of sulfate attack. In case of acid attack, the cement stone itself (calcium silicate hydrate) is also degraded by the aggressive liquids. More details on these degradation processes will be given in Chapter 5.

Whether it is due to the pore structure and transport mechanisms or due to nature and amount of hydration products, it is clear that the cement matrix plays a crucial role concerning the durability of concrete. As the matrix is the result of the interaction between cement and water (and possibly also additions and admixtures), it is not surprising that composition

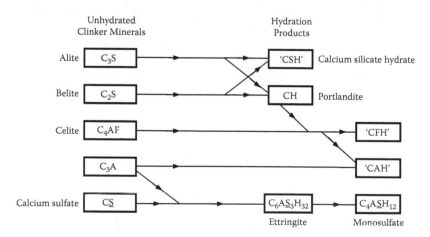

Figure 1.7 Hydration of Portland cement.

parameters like cement content (C), water content (W), and their mass ratio called the water/cement ratio (W/C) are key parameters for the evaluation of the durability of concrete. The role of these parameters will be explained in the following paragraphs. Some critical remarks however will also be given further in this chapter.

1.4.2 Water/cement ratio

Considering that no aggressive substances are mixed within the concrete itself, the evolution of degradation mechanisms is largely dependent on the penetrability of the concrete for aggressive substances coming from the environment. In other words, the permeability of the concrete will play an important role. For cement pastes with similar degrees of hydration, a higher water/cement ratio will lead to a higher porosity and a higher permeability. While porosity could be considered to linearly depend on the water/cement ratio, permeability evolves in a highly non-linear way, as shown in Figure 1.8.

For lower water/cement ratios (0.3 to 0.5), the inclination of the curve is significantly lower than for higher water/cement ratios (0.6 to 0.7). A decrease of the water/cement ratio from 0.7 to 0.3 decreases the permeability of the paste by at least two orders of magnitude. This drastic influence of the water/cement ratio on permeability can also be found for concrete. For a water/cement ratio of 0.75, a typical concrete permeability is in the order of 10^{-10} m/s, while it is 10^{-11} to 10^{-12} m/s for a water/cement ratio of 0.45 (Neville 1995).

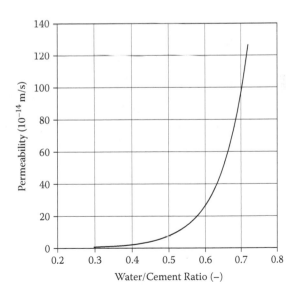

Figure 1.8 Relation between permeability and water/cement ratio of cement paste.

Figure 1.8 illustrates the major importance of the water/cement ratio for the permeability and thus for the durability of paste, mortar, and concrete. For a water/cement ratio higher than 0.5, a pore structure is obtained with a higher degree of percolation. The more the pore structure percolates, the faster aggressive substances can enter the material. Avoiding a percolating pore structure should be considered as a necessary condition in order to obtain a durable concrete. In this respect, concrete with a water/cement ratio larger than 0.6 can hardly be named durable. Durable concrete requires a limited water/cement ratio.

1.4.3 Cement content

As such, the cement content is not a decisive parameter for the durability of concrete. It is clear that the cement content does have an influence for instance on the amount of chlorides that can be chemically bound in concrete or the amount of carbon dioxide that can be buffered by the portlandite. However, the influence of cement content will be less determining than the influence of the water/cement ratio.

As the durability (and also the strength) of concrete depends on the properties of the hydrated cement matrix, the cement concentration within this matrix surely has an influence. On the other hand, the required cement matrix volume is dependent on the maximum particle size of the aggregates. For a larger aggregate size, the needed matrix volume to 'glue' the particles is lower. As a result, the cement content per cubic meter of concrete should be studied in relation to the maximum particle size applied in the concrete. This interrelation was well understood in the former British Standard BS 5328:Part 1:1991, prescribing a correction of the required cement content per cubic meter of concrete depending on the maximum particle size, as shown in Table 1.1.

ACI Committee 207 (1970) also agreed on a correction of the cement content depending on the applied maximum particle size. However, based on experimental research, they also concluded that a certain optimal maximum particle size can be defined above which a further reduction of the cement content is no longer acceptable. Above this optimal value, an increasing cement content will be needed to obtain comparable concrete properties.

Table 1.1 Correction on required cement content (BS 5328:part 1:1991)

Maximum particle size (mm)	Correction on required cement content (kg/m³)
10	+40
20	0
40	−30
80	−70

1.4.4 Cement type

In order to characterize the durability of concrete, it is not sufficient to know the water/cement ratio and the cement content. The cement type is also very important with regard to concrete durability. As an example, concrete based on blast furnace slag cement or containing silica fume will provide a better resistance against chloride penetration. In the case where the slag content of the binder is at least 60%, the chloride diffusion coefficient is reduced by a factor of ten in comparison with Portland cement-based concrete with a similar water/cement ratio of 0.5 (Neville 1995). On the other hand, blast furnace slag-based binders with such high slag contents can be more sensitive to carbonation. Carbon dioxide will not be sufficiently bound in the concrete cover and the pore system will not be blocked with calcium carbonate, in contradiction to what happens in Portland systems. This illustrates the difficulties involved in defining durability requirements solely based on cement content and water/cement ratio. A similar remark can be formulated towards strength as a governing parameter.

Besides chloride penetration and carbonation, the nature of the cementitious materials is also relevant for other degradation mechanisms. Furthermore, the selection of cement type can be influenced by requirements concerning early age strength development (depending on the casting temperature), heat of hydration, shrinkage, etc. Puzzolan or other additions can be very advantageous in some cases. On the other hand, it should also be mentioned that in case of puzzolan or other slowly reacting additions, the concrete is more sensitive to deficient curing. Casting and curing conditions generally also influence concrete durability, as will be explained further on.

With the example of chloride penetration and carbonation, the importance of the right choice of cement type or binder system is illustrated. Not every cement type is adequate for every application. Well-known is the choice of a high sulphate resistant cement or a low alkali cement in some specific conditions. A durable concrete requires a motivated choice of cement type or binder system, duly considering the cement chemistry (Hewlett 1988).

1.4.5 Strength

Durability (and sustainability) of concrete structures nowadays is an important issue. The due attention to durability is quite in contrast with earlier ideas that only concrete strength is of importance, implicitly considering that a strong concrete will automatically perform as a durable concrete. To illustrate the older idea that concrete is inherently durable, reference is made to the British Standard Code of Practice CP 114 valid in 1948: "No structural maintenance should be necessary for dense concrete constructed in accordance with this code" (Neville 1995). It is clear that the sole or at least main attention to concrete strength has led to serious durability issues in the past.

The dubious role of strength as a durability indicator can be further illustrated by studying some historical information. The general applicability of concrete as a construction material significantly increased in the twentieth century thanks to extensive laboratory research coupled with increasing practical experience. New cements have been developed with increased chemical resistance. It was gradually realized that the quality of the concrete determines the resistance against degradation. In this respect, quality of the concrete was often interpreted in a too limited way as the strength of the concrete. However, a strong concrete is not automatically a durable concrete. Some situations can be found where a stronger concrete can show more degradation than a weaker concrete, when the latter is based on a more resistant binder system. The correlation between strength and durability might hold more or less for concretes based on the same cement type, but it certainly does not hold when comparing concretes with different binder systems.

Some interesting additional arguments can be obtained by studying the historical evolution of cement properties. Typically, cement strengths have increased during the last century (Neville 1997, De Schutter 2001). Due to the increase in cement strength, a concrete with characteristic cube strength of 32.5 N/mm^2 could be obtained in 1984 with a water/cement ratio of 0.57, while in 1970 a water/cement ratio of 0.50 was needed. Because of the required workability level, the water content per cubic meter of concrete remained about the same, while the cement content could be significantly reduced maintaining the same concrete strength level. Neville (1997) concludes that the cement content per cubic meter of concrete could be reduced by 60 to 100 kg, leading to an increase in water/cement ratio of 0.09 to 0.13. It is clear that, in spite of a similar strength, the concrete will show an increased porosity and permeability, and thus will be more prone to degradation mechanisms like carbonation and chloride penetration. Furthermore, due to the faster strength development at early age, formworks could be removed faster. This further increases the risk of inadequate curing conditions, which negatively influences the final quality of the concrete, including durability performance (see next section).

1.4.6 Casting and curing

The durability of concrete structures not only depends on the intrinsic durability performance of the concrete mix, but to a large extent also on the real casting and curing conditions. The difference is sometimes referred to as the difference between 'labcrete' (concrete cast and cured in perfect laboratory conditions) and 'realcrete' (concrete cast and cured in real conditions, on site). Three major aspects are important in this context: compaction, cover thickness, and curing.

Traditional concrete needs to be compacted when poured into the formwork. Uncompacted concrete contains a high percentage of voids, leading to a strength reduction and to increased transport properties (higher permeability). Compaction of concrete on construction sites is traditionally performed by means of vibration needles or pokers. In precast plants, vibrating tables or formwork vibration can be applied as well as other techniques. Nowadays self-compacting or self-consolidating concrete (SCC) is also available on the market, making compaction by external vibration obsolete (De Schutter et al. 2008). Modern SCC was developed in the 1980s in Japan, in order to overcome durability problems related to badly vibrated traditional concrete. For a more detailed discussion on the intrinsic durability behaviour of SCC, reference is made to literature (De Schutter and Audenaert 2007).

During casting, due attention has to be given to the cover thickness. Positioning of formwork and reinforcement cages should be done with utmost care in order to avoid too low values of the cover thickness. Prescribed cover thickness values should be considered as minimum values and should be carefully respected in practice by appropriate use of spacers. The cover thickness is very important in order to avoid reinforcement corrosion due to carbonation or chloride penetration. The concrete within the cover zone, the so-called 'covercrete', provides a barrier against carbon dioxide and chloride ions. The higher the cover thickness, the longer it takes before the carbonation depth or a critical chloride content reaches the steel reinforcement (see Chapter 5). It is important to realize that the time it takes for the carbonation front to reach the steel reinforcement is not linearly proportional to the cover thickness. The carbonation depth is typically more or less proportional to the square root of time with some retardation at later age (Figure 1.9).

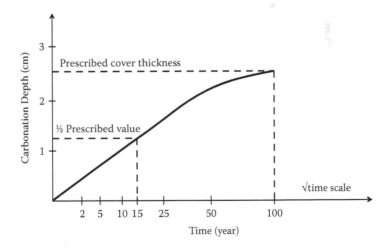

Figure 1.9 Effect of reduced cover thickness on service life in case of carbonation.

To illustrate the effect of a reduced cover thickness, suppose that the end of service life is reached when the carbonation depth reaches the steel reinforcement. According to the evolution shown in Figure 1.9, a concrete cover thickness of 2.5 cm is required to reach a service life of one hundred years. In case the real cover, due to inadequate positioning of formwork and reinforcement cage, is reduced to half the prescribed value, the service life is tremendously reduced to only fifteen years (and not to fifty years, which would be half the targeted service life!). This numerical example clearly illustrates the importance of cover thickness. An effective quality control system should be defined on site in order to make sure that the prescribed cover thickness is strictly respected.

Freshly cast concrete should also be properly cured in order to make sure that the cement hydration proceeds in a good way and that no early age cracks occur due to loss of water. Depending on some important parameters (such as the hardening rate of the concrete and temperature and humidity of the environment), the curing period should be appropriately determined. It is not the intention of this text to outline different curing methods. It will only be explained here what the effect of improper curing could be on the final durability of the concrete element.

A first consequence of bad curing could be a reduction of the concrete quality because early loss of water might lead to a premature end of the hydration process which can only partly be recovered later on by rewetting. The effect of this reduced concrete quality can be compared with the effect of an increased water/cement ratio, as shown in Figure 1.10. The cover

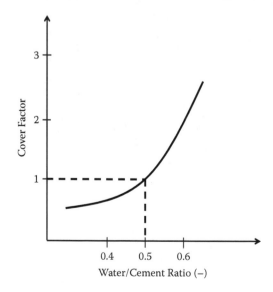

Figure 1.10 Influence of concrete quality on the cover thickness.

factor, shown in the vertical axis, refers to the proportional increase in cover thickness needed to compensate for the reduced concrete quality as indicated by the increasing water/cement ratio. A lower concrete quality (due to increased water/cement ratio or improper curing) will quickly lead to a significantly higher required cover thickness in order to reach the target service life.

A second consequence of bad curing can be the formation of plastic or drying shrinkage cracks at early age. As the loss of water is accompanied by a volume reduction of the concrete, shrinkage cracks could occur in the drying concrete surface. As a result, the hardened concrete element will contain some small shrinkage cracks, which will further influence the transport properties. Through the cracks, aggressive liquids and gasses can easily penetrate the concrete and reach the reinforcement much faster than expected. As an example, the effect of small shrinkage cracks on the penetration of chloride ions is schematically illustrated in Figure 1.11, as experimentally obtained on laboratory samples with crack width 0.3 mm and crack depth 20 mm exposed to chloride migration tests. Due to this effect, chloride ions can very quickly initiate corrosion of the reinforcing bars. The penetration of carbon dioxide is somewhat less sensitive to the existence of very small cracks below 0.3 mm. Although the above explanation is linked to the occurrence of shrinkage cracks due to lack of curing, a similar influence is obtained in case of early age thermal cracking due to heat of hydration (see also Chapter 4).

1.5 PRACTICAL DURABILITY APPROACH

1.5.1 Typical code provisions

Because the previously discussed parameters water/cement ratio and cement content have been considered important durability parameters for a long time, it is no surprise to see that typical code provisions related to durability of concrete are often based on them. In Europe, the standard EN 206-1 'Concrete—Part 1: Specification, performance, production and conformity' departs from the notion of exposure classes, designated with a capital letter X, followed by another letter depending on the specific degradation mechanism to be considered: C for carbonation, D for de-icing salts, S for sea water, F for frost, and A for chemically aggressive environment. To this letter combination, a number is added, which in most cases is linked to specific humidity conditions. In total, 18 exposure classes have been defined, as listed in Table 1.2.

Depending on the environment, several degradation mechanisms can occur in parallel. Therefore, it is necessary to select all relevant exposure classes for the considered application. Finally, the concrete composition will have to be designed considering the most severe exposure class.

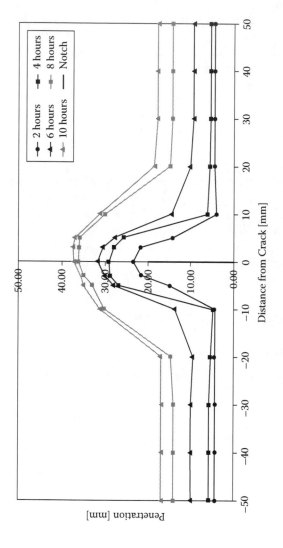

Figure 1.11 Influence of cracks (width 0.3 mm, depth 20 mm) on accelerated chloride migration in laboratory test.

Table 1.2 Exposure classes defined in European Standard EN 206-1:2000

Exposure classes	
No risk of corrosion or attack	
X0	
Corrosion induced by carbonation	
XC1	Dry or permanently wet
XC2	Wet, rarely dry
XC3	Moderate humidity
XC4	Cyclic wet and dry
Corrosion induced by chlorides other than from sea water	
XD1	Moderate humidity
XD2	Wet, rarely dry
XD3	Cyclic wet and dry
Corrosion induced by chlorides from sea water	
XS1	Exposed to airborne salt but not in direct contact with sea water
XS2	Permanently submerged
XS3	Tidal, splash, and spray zones
Freeze/thaw attack with or without de-icing salts	
XF1	Moderate water saturation without de-icing agents
XF2	Moderate water saturation with de-icing agents
XF3	High water saturation without de-icing agents
XF4	High water saturation with de-icing agents or sea water
Chemical attack	
XA1	Slightly aggressive chemical environment
XA2	Moderately aggressive chemical environment
XA3	Highly aggressive chemical environment

The durability requirements linked to the listed exposure classes are expressed in terms of a maximum allowable water/cement ratio and a minimum needed cement content. In an indicative way, also strength classes are mentioned. All requirements in terms of water/cement ratio, cement content and strength are locally defined in each European member state. The code provisions of the 'place of use' should be followed when producing concrete elements in Europe.

While the durability provisions of the European Standard EN 206-1:2000 were summarized in the previous paragraphs as an example, it can be mentioned that a similar approach is followed in many other international standards such as the North American Standard ACI 318:2011, the Canadian Standard CSA A23.1:2009, the Australian Standard AS 3600:2001, and the Indian Standard IS 456-2000. Some exposure classes are defined referring to the anticipated severity of the environment of the concrete element. The exposure classes are subdivided depending on humidity conditions. According to the exposure (sub)class, limiting values are specified for water/cement ratio and compressive strength in the North American, Canadian and Australian standards, while a minimum cement content is also required in the European standard. Additional requirements might be added in some cases, such as high sulfate resisting cement (EN 206-1). A comprehensive overview is given by Kulkarni (2009).

1.5.2 Some critical reflection on typical code provisions

Each of the parameters considered in typical code provisions (water/cement ratio, cement content, concrete strength) can be questioned (Figure 1.12). It is clear (although not always well-known in concrete practice) that concrete strength alone cannot ensure durable behaviour (Neville 1997). Furthermore, the minimum cement content requirement can be questioned (Wasserman et al. 2008). Even the water/cement ratio, which is generally believed to be the most important parameter governing the durability behaviour, should be applied with caution. "We should remember the limitations on its interpretation", as stated by Neville (1999).

The definition of the water/cement ratio became more complicated after the introduction of supplementary cementitious materials like fly ash, silica fume, and blast furnace slag. In order to more accurately estimate the effect of these various cementitious materials, the k-value concept has been implemented in the European Standard EN 206-1. However, major discussions

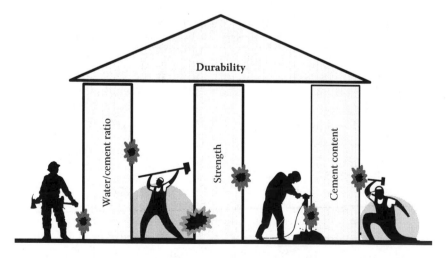

Figure 1.12 Criticism on typical code provisions.

arise in concrete practice about the actual k-values, and about the validity of this k-value concept. Different European member states have different opinions and prescriptions in this respect, as mentioned in the various national application documents linked to EN 206-1.

It is worth mentioning that k-values are defined based on compressive strength results, while the effect of the supplementary cementitious materials on durability behaviour can be quite different. This fundamentally undermines the practical application of the k-value concept. Each durability application of this concept should be done with care!

Because the basic parameters (water/cement ratio, cement content and concrete strength) can be criticised as governing parameters for the durability of concrete, some attempts are made to ensure durability, e.g. by water absorption by immersion. This is also implemented in the EN 206-1, as a possible additional requirement. However, it is to be mentioned that water absorption by immersion has to be applied with caution (De Schutter and Audenaert 2004).

1.5.3 Equivalent concrete performance concept (ECPC)

One possible way to overcome everlasting discussions about k-values is to check the equivalency of concrete performance as described within the European code EN 206-1. As an alternative to the durability requirements given as functions of water/cement ratio and cement content, the durability performance of a specific non-traditional concrete (containing an alternative binder system) can be demonstrated experimentally in comparison with a generally accepted "deemed to satisfy" concrete, which is fully in agreement with the classical durability requirements. This approach is called the Equivalent Concrete Performance Concept (ECPC).

As described in the European code EN 206-1, paragraph 5.2.5.3, "The equivalent concrete performance concept permits amendments to the requirements in this standard for minimum cement content and maximum water/cement ratio when a combination of a specific addition and a specific cement is used, for which the manufacturing source and characteristics of each are clearly defined and documented". The (informative) Annexe E of EN 206-1 provides further details of the equivalent concrete performance concept. "Testing should show that the performance of the concrete containing the addition should be at least equivalent to that of the reference concrete." Furthermore, some general guidance is given in Annexe E, which can form the basis for an actual testing and acceptance procedure.

Unfortunately, European Standard 206-1 does not give a detailed method for the practical implementation of the ECPC concept. In Belgium, a standard has been developed for the practical implementation of the ECPC concept, enabling the evaluation and attestation of concrete with non-standard cement types and/or puzzolanic additions. This standard, NBN B15-100, considers two important terms: 'general suitability' and 'specific suitability'. General suitability refers to the fact that a constituent material can be successfully used within concrete. Materials having a CE-mark are considered to be generally suitable. For non-certified materials, the general suitability first has to be evaluated experimentally (e.g. composition, alkali content, chloride content, mechanical properties, stability, heat of hydration, setting, loss on ignition, residue, sulphate content).

The most important part of the standard NBN B15-100 deals with specific suitability. This refers to the suitability of a specific mix for a specific field of application and exposure conditions. The specific suitability has to be checked by means of Initial Type Testing (ITT) on real concrete mixes. The evaluation of the durability is based on comparative laboratory tests, comparing the non-traditional concrete composition with standard solutions accepted by EN 206-1. For more details on this method, reference is made to De Schutter (2009).

1.5.4 Durability indicators

Although the ECPC concept provides an excellent opportunity to evaluate the durability of concrete based on new binder types, the main discussion point is the definition of well-accepted reference concrete for the considered exposure classes. This is inherent to the fact that the ECPC concept is a comparative method, comparing durability performance with accepted deemed-to-satisfy solutions. A more sound evaluation of durability performance could be based on absolute performance criteria. Although we are still far from defining absolute durability performance criteria for completed concrete structures or concrete elements, considerable progress is being made following the approach of so-called durability indicators (Baroghel-Bouny et al. 2009), combined with fundamental models.

A list of general durability indicators, which are relevant to many degradation processes, can be defined including calcium hydroxide content, water accessible porosity, chloride ion diffusion coefficient, and gas or water permeability. In order to study the durability performance with regard to some specific degradation mechanisms, some specific durability indicators can be considered, e.g. relevant to freeze-thaw damage or alkali-silica reaction. Some complementary parameters can also be needed as input data for the fundamental prediction models, e.g. the chloride concentration at the concrete surface. Based on the durability indicators, durability classes can be proposed, ranging from very low to very high potential durability as illustrated in Figure 1.13. More detailed examples can be found in literature (Baroghel-Bouny et al. 2009).

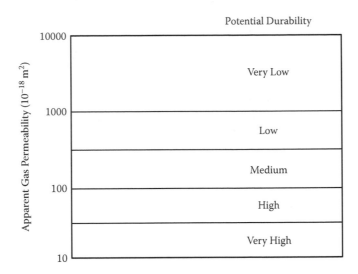

Figure 1.13 Example of durability classification based on a durability indicator.

The approach based on durability indicators seems to offer more freedom to achieve a required durability performance. Further developments, however, are needed in order to come to a generally accepted method which can be easily applied.

1.6 MORE ADVANCED DURABILITY DESIGN

A more advanced durability design can be based on reliability theory, as applied to mechanical or structural design. Degradation mechanisms can be introduced into the reliability analysis by explicitly considering time, yielding a time-dependent failure probability.

In a simplified approach, a load-action S and a resistance R can be considered, which can be any stochastic variable expressed in arbitrary units. The only requirement is that S and R can be directly compared. Failure will occur when the resistance R becomes smaller than the load-action S, or R < S. For a time-independent situation, the probability of failure is then defined as the probability that R < S:

$$P_f = P\,[R < S] \tag{1.1}$$

Considering degradation mechanisms, it is clear that both the resistance R and the load-action S can be time-dependent stochastic variables, as illustrated in Figure 1.14a. This will lead to a time-dependent failure probability:

$$P_f(t) = P\,[R(\tau) < S(\tau)] \qquad \text{for all } \tau \le t \tag{1.2}$$

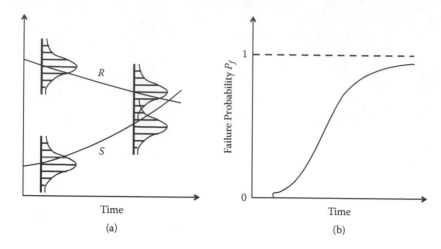

Figure 1.14 Time-dependent probability of failure.

The time-dependent failure probability is illustrated in Figure 1.14b. The failure probability curve could start with an initial non-zero value which represents the failure probability when loading the structure for the first time. Afterward, the failure probability is increasing and will eventually reach a value of one.

The situation of Figure 1.14 resembles the classical case of a decreasing resistance R and an increasing load-action S. At early age of a concrete structure, it is also common to have a somewhat increasing resistance R, due to further (slow) hydration of the cement. An increasing resistance R as a function of time is also referred to as 'negative aging'.

From a statistical point of view, the function $P_f(t)$ has all the characteristics of a cumulative distribution function (CDF). When the service life L is defined in such a way that the event "L<t" has the same meaning as the event "the structure fails within the time interval (0, t)", the following equation can be written, with F_L the CDF of the service life:

$$F_L(t) = P\,[L < t] = P_f(t) \tag{1.3}$$

When the CDF of the service life is known, the probability density function (PDF) is obtained by:

$$f_L(t) = d\,F_L(t)/dt \tag{1.4}$$

and consequently:

$$f_L(t)\,dt = P\,[t < L < t + dt] \tag{1.5}$$

Reaching the service life within the time interval (t, t + dt) implies that R < S within (t, t + dt) and that R > S in (0, t). Indeed, if this latter condition is not fulfilled, the service life would already have been reached before the interval (t, t + dt). Consequently, it can be written that:

$$f_L(t) \, dt = P \, [R < S \text{ in } (t, t + dt) \text{ and } R > S \text{ in } (0, t)] \tag{1.6}$$

The PDF of the service life L can be considered as the failure probability per time unit or the failure rate. In this respect, the terminology 'unconditional failure rate' is also applied, in contrast to the 'conditional failure rate' or 'hazard function' r(t). This hazard function

$$r(t) \, dt = P \, [R < S \text{ in } (t, t + dt) \mid R > S \text{ in } (0, t)] \tag{1.7}$$

is the probability that a system or component, which was functioning in the time interval (0, t), will fail in the following elementary time interval dt. The hazard function gives an indication whether it is becoming more probable for an element to fail with aging.

It can be shown that the following relations exist between r(t) and the distribution function of the service life L:

$$r(t) = f_L(t)/(1 - F_L(t)) \tag{1.8}$$

$$F_L(t) = 1 - \exp \left[- \int_0^t r \, (\tau) \, d\tau \right] \tag{1.9}$$

Figure 1.15 shows some typical cases for $F_L(t)$, $f_L(t)$ and $r(t)$. In the first case (a), the hazard function r(t) is constant. The conditional failure rate is constant in time. Applying equation (7), this leads to an exponential distribution of the service life:

$$F_L(t) = 1 - \exp \, (-\lambda t) \qquad (t \geq 0, \, \lambda = \text{constant}) \tag{1.10}$$

In case (b) of Figure 1.15, the hazard function r(t) decreases with time due to ongoing hydration and increased strength of the concrete. In case (c), an increasing function of r(t) is shown, which is typical in case of degradation, ageing or damage. In reality, a combination of the different situations can occur, including a first period with decreasing hazard function, a second period with constant values, and a third period with increasing hazard function. The resulting overall evolution of the hazard function is then typically as shown in Figure 1.16.

Instead of considering the service life as a stochastic variable, a more classical reliability analysis typically starts from a well-defined reference

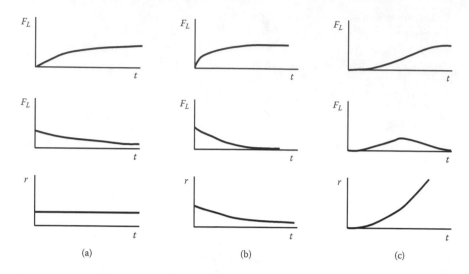

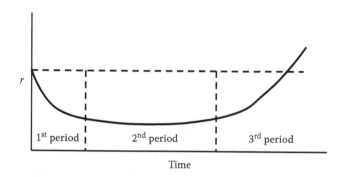

Figure 1.15 Different service life distribution types.

Figure 1.16 Example of a hazard function for a real structure.

period (e.g. 100 years). The failure probability P_f or the reliability index β is then calculated considering the time-dependency of resistance R and load-action S. In a further simplification of the durability design, an approach based on partial safety factors can be defined analogous with structural design codes, avoiding the need to perform detailed statistical calculations based on complete distribution functions. For more details on a probabilistic durability design, reference is made to literature (Sarja and Vesikari 1996).

An often cited probabilistic durability design model is the Duracrete Model, developed within a European Research Project (Brite-EuRam 1998, 1999). With this model, the service life of a structure is studied by considering the stochastic nature of degradation mechanisms like chloride-induced corrosion, etc. The difficulty with these kinds of stochastic models is to select the distribution function of the variables and to define the appropriate parameter values. The modelled service life might be sensitive to these choices. Further research is certainly needed.

REFERENCES

ACI Committee 207 (1970) 'Mass concrete for dams and other massive structures', *ACI Journal*, April, pp. 273–309.

Baroghel-Bouny V., Nguyen T.Q., and Dangla P. (2009) 'Assessment and prediction of RC structure service life by means of durability indicators and physical/chemical models', *Cement and Concrete Composites*, 31, pp. 522–534.

Brite-EuRam (1998) 'Modelling of Degradation', Report R4-5, Brussels: BE-1347/TG7/Report R4-5, Project No. BE95-1347, DuraCrete: 'Probabilistic Performance Based Durability Design of Concrete Structures'.

Brite-EuRam (1999) 'General Guidelines for Durability Design and Redesign', Brussels: BE-1347/TG7/Report R14, Project No. BE95-1347, DuraCrete: 'Probabilistic Performance Based Durability Design of Concrete Structures'.

De Schutter G. (2001) 'Evolution of cement properties in Belgium since 1950', *Magazine of Concrete Research*, 53, 5, pp. 291–299.

De Schutter G. and Audenaert K. (2004) 'Evaluation of water absorption of concrete as a measure for resistance against carbonation and chloride migration', *Materials and Structures*, Vol. 37, No.273, pp. 591–596.

De Schutter G. and Audenaert K. (eds.) (2007) 'Durability of Self-Compacting Concrete', State-of-the-art report of RILEM Technical Committee 205-DSC, RILEM Report 38, RILEM Publications S.A.R.L., ISBN 978-2-35158-048-6, pp.185.

De Schutter G., Bartos P., Domone P. and Gibbs J. (2008) 'Self-Compacting Concrete', Whittles Publishing, Caithness, UK, CRC Press, Taylor & Francis Group, Boca Raton, USA, ISBN 978-1904445-30-2, USA ISBN 978-1-4200-6833-7, p. 296.

De Schutter G. (2009) 'How to evaluate equivalent concrete performance following EN 206-1 ? The Belgian Approach', Proceedings of the Second International RILEM Workshop "Concrete Durability and Service Life Planning—ConcreteLife '09", Haifa (Israel), 7–9 September 2009, ed. Konstantin Kovler, Published by RILEM Publications (Bagneux, France), PRO 66, pp. 1–7, ISBN 978-2-35158-074-5.

Hewlett P.C. (1988) 'Lea's chemistry of cement and concrete', Fourth Edition, J. Wiley & Sons.

Kulkarni V.R. (2009) 'Exposure classes for designing durable concrete', *The Indian Concrete Journal*, March, pp. 23–43.

Neville A.M. (1995) 'Properties of Concrete', Fourth Edition, Longman Group Limited.

Neville A.M. (1997) 'Maintenance and durability of structures', *Concrete International*, November, pp. 52–56.

Neville A.M. (1999) 'How useful is the water-cement ratio', *Concrete International*, September, pp. 69–70

Sarja A. and Vesikari, E. (eds.) (1996) *'Durability design of concrete structures'*, RILEM Report 14, E&FN Spon, London.

Wasserman R., Katz A. and Bentur A. (2008) 'Minimum cement content requirements: a must or a myth', Materials and Structures, DOI 10.1617/s11527-008-9436-0.

Chapter 2

Inappropriate design

2.1 INTRODUCTION

Completing a structure starts with an appropriate design, based on structural requirements. As already mentioned in Chapter 1 (Figure 1.5), the design stage includes dimensioning, detailing, and material selection based on properly defined environmental conditions, duly considering both mechanical loading and potential degradation processes. Structural engineers are well-trained in estimating the external loading, as well as in determining the resulting internal forces and stresses in structural elements. To do so, many tools can be considered, going from hand calculations to estimate order of magnitudes to very complex computer models giving accurate numbers for complex geometries under combined loading effects.

The fundamental theories behind structural behaviour are well known and can be translated into sound mathematical language. In this way, the calculation of the internal forces in a structural element can be obtained by solving fourth order partial differential equations linking deformations with external loading. Different material models can be considered to finally calculate stress distributions, going from simple linear elastic behaviour to more complex behaviour including visco-elasticity, plasticity, cracking, and anisotropy.

Dimensioning the structure can finally be based on the concept of allowable stress levels (as typically done in elastic design methods) or on the definition of (partial) safety factors relative to the failure condition of the structural element (as typically done in ultimate limit state design methods, ULS). Especially for slender structures, due attention should also be paid to allowable deformations (as defined in the so-called serviceability limit state, SLS). These concepts are well documented in literature or in design codes, and will not be dealt with in the framework of this book on damage to concrete structures.

Although design engineers are well acquainted with these concepts, due to the complexity of the structure, often in combination with uncertainties regarding the external loading, things can go wrong. Fortunately, structural failure does not happen very often, making engineering structures highly ranked within the list of most reliable man-made objects (more reliable than chips and computers). However, in the event of structural failure, the consequences can be very high, including human lives.

2.2 INAPPROPRIATE DIMENSIONS AND DETAILING

Allen (1979) collected information on 188 cases of failing concrete structures, 29 resulting in collapse and 118 in unacceptable serviceability issues such as deterioration, excessive cracking, spalling, deflection or settlement. It was concluded that about half of the errors were due to inappropriate design, and the other half due to faulty construction (see Chapter 3). Within the group of collapsed structures, inadequate design (e.g. of formwork or temporary bracing) or inadequate detailing was responsible for most of the cases. Many of the serviceability problems within the other group were also due to improper design related to deflection, temperature effects, shrinkage, and creep. The conclusion that about half of the structural failures are caused by errors in design or lack of design was also reached by Ioniṭă et al. (2009) within their analysis. They attribute another quarter of the failures to errors on the construction site. Human error is a major cause of structural failures, independent from the type of material. Human errors can be due to ignorance (*the designer is not aware of the problem*), due to carelessness (*the designer is aware of the problem, but does not care about it*), or by intention (*the designer is aware of the problem, even cares about it, but takes the risk to oversimplify due to time pressure or lack of budget*) (Kaminetsky 1991). Ignorance and insufficient knowledge are reported to be the main cause of human error, representing about two thirds of the cases (Matousek and Schneider 1976, Kaminetsky 1991, Natarajan 2007).

To make mistakes is very human, as nicely illustrated for the case of engineers by Petroski (1982). Human error cannot be totally avoided. On the other hand, we cannot just sit back and give the fatalistic *'that-is-life'* excuse. Efforts should be made to reduce the number of human errors within structural design. As mentioned by Kaminetsky (1991), *'training of engineers and control in the design phase should have high priority, since the most frequent errors are made in this phase'*. Normal care and additional control are considered accurate measures to avoid almost all human errors in structural design. Only about 10% of structural failures and errors seem to be reasonably unavoidable (Kaminetsky, 1991). Structural collapse due to inappropriate dimensions or detailing will be illustrated hereafter by referring to three cases: the collapse of the Melle Bridge in Belgium, the sinking of the Sleipner A offshore platform in Norway 1991, and the tumbling collapse of a 13-storey apartment building in Shanghai, China 2009.

2.2.1 Collapse of the Melle Bridge, Belgium, 1991

In 1991, the Melle Bridge, Belgium (Figure 2.1) suddenly collapsed while loaded by one truck. As a consequence, the truck (carrying gasoline) exploded, and the driver was killed. The reason for the collapse was not immediately clear, because the bridge, built in the 1960s, was designed to carry much higher loading than a mere gasoline truck. Detailed investigation revealed the fundamental reason, linked with inappropriate detailing. The Melle Bridge was a three-span prestressed bridge, as schematically shown in Figure 2.2. The main span was bridging the canal, while the side spans were very short, only bridging the walkways along the canal. At the abutments, some tensile pendulum pillars were provided, as shown in the details in Figure 2.2, and in Figure 2.3. These pendulum pillars were prestressed by means of prestressing wires. In order to enable the pendulum movement, some hinges were provided in these pillars, as schematically shown in the detail of Figure 2.2.

The hinge was provided by reducing the concrete section to one third of the full section, leaving only about 5 cm of concrete with centrally positioned prestressing wires. The concrete cover thickness protecting the wires thus was locally reduced to less than 20 mm. Furthermore, due to the hinging action, this section contained microcracking, which further accelerated the penetration of chloride containing de-icing salts towards the prestressing wires, which are very sensitive to chloride-induced corrosion. As a result, the prestressing wires were strongly corroded, as can be seen in Figure 2.4, leading to a severely reduced load carrying capacity of the tensile pendulum pillars. After about twenty-five years of service, progressive failure occurred while the bridge was loaded by the gasoline truck. One of

Figure 2.1 Collapse of the Melle Bridge, Belgium, 1991.

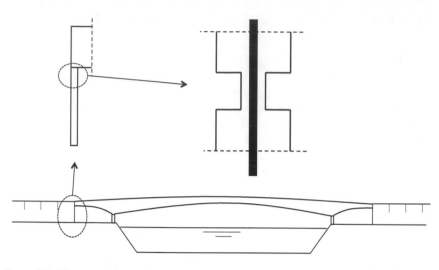

Figure 2.2 Schematic view of the Melle Bridge, and some details (tensile pendulum pillar and hinge).

Figure 2.3 Picture of the tensile pendulum pillars (after failure).

the pendulum pillars failed, leading to a load increase in the other pendulums, which in their turn failed. Due to the failure of the tensile pendulum pillars, a moment redistribution occurred in the main span of the bridge, leading to values higher than the design value. As a consequence, the main span also failed, ending in the complete collapse of the Melle Bridge.

In summary, a design detail (the hinges) indirectly caused the collapse of the bridge. Unfortunately, the bridge seemed not very robust. Although the bridge was hyperstatic (statically indeterminate), failure of the pendulums immediately provoked failure in the main span, giving a nice but

Figure 2.4 View of failed hinge of tensile pendulum pillar (with corroded prestressing wires).

unfortunate example of progressive collapse. In a more robust bridge design (Ioniță et al. 2009), the consequences of a badly conceived detail would have been less dramatic.

2.2.2 Sinking of the Sleipner A offshore platform, Norway, 1991

One of the most impressive failures of a man-made structure was the sinking of the Sleipner A offshore platform in Norway in 1991 (Collins et al. 2000). The Sleipner A platform was a concrete base structure consisting of 24 cells and 4 shafts, 110 meters high. On August 23, 1991, as the platform was gradually positioned deeper into the water of a fjord near Stavanger in order to mount the platform deck on top of the 4 shafts, a sudden leak occurred in one of the cells. The entire platform sank to the sea bottom, causing a seismic event of magnitude 3.0 on the Richter scale, and only leaving a pile of concrete rubble in the deep fjord. The economic losses were enormous, going into hundreds of millions of American dollars.

At the request of the Norwegian structural engineering firm responsible for the design of the Sleipner A platform, Professors Collins and Vecchio of the University of Toronto, Canada, thoroughly studied the reasons for the dramatic loss of the platform (Collins et al. 2000). They came to the following conclusion: "*While the exterior walls of the cells were circular, with a radius of 12 meters, the interior walls, which separated the cells, were straight. At the interior section points of these interior walls, a small triangular void called a tricell was formed. There were a total of 32 such tricells. Because these tricells had openings at the top, they filled with water once the top of the cells were submerged. Therefore, the walls of the tricells had to resist a substantial hydrostatic pressure. The loss of the structure was attributed to the failure of the wall of one of these tricells".*

Reading this conclusion, one might think that the designers made a substantial error. Hydrostatic pressure is a well-known loading case, and the design of a structural element withstanding this pressure should have been straightforward. Or not? One interesting consideration related to the specific type of concrete structure considered here, is formulated by Collins and Vecchio as follows: *"However, unlike the situation for a typical land-based structure, the designer does not have the option of greatly increasing the wall thickness to ensure a very conservative design. If the walls are too thick, the structure will not float, or will not be hydrostatically stable during the tow to the field. These severe constraints mean that for these weight-sensitive structures rather low factors of safety are employed. As a consequence, great care is required in all aspects of design and construction".*

This last sentence might really point to the main issue concerning the design of the Sleipner A platform. As previously mentioned in this chapter, additional control can be very helpful to avoid many human errors in structural design. Finite element simulations with commercially available software were performed in order to calculate internal forces and stresses within the cell walls, including the tricells. As mentioned by Collins and Vecchio, *"The tricell wall that failed did not contain stirrups because the global finite element analysis performed as part of the design seriously underestimated the magnitude of the shear at the ends of the walls, while the sectional design procedures used seriously overestimated the beneficial effects of axial compression on the shear strength of the wall".*

It is always easy to give comments afterwards, once the failure has occurred. Nevertheless, Collins and Vecchio are very clear when legitimately formulating one of their general recommendations, as one of the lessons learnt from this remarkable failure: *"... no matter how complex the structure or how sophisticated the computer software, it is always possible to obtain most of the important design parameters by relatively simple hand calculations. Such calculations should always be done, both to check the computer results and to improve the engineers' understanding of the critical design issues".*

It is worthwhile mentioning that afterwards, refined finite element analysis has been capable of predicting the water depth at which failure of the Sleipner A platform occurred. This kind of sophisticated computer model is very valuable indeed, but common engineering sense is always needed to verify orders of magnitude.

2.2.3 Collapse of a 13-storey apartment building in Shanghai, China 2009

On June 27, 2009, a nearly finished 13-storey apartment building in Shanghai just tumbled over, killing one worker (see Figure 2.5). As in many collapses,

Figure 2.5 Tumbling collapse of apartment building in Shanghai, 2009 (Internet photo).

Figure 2.6 Detail of the foundation piles (Internet photo).

a combination of several aspects led to this disaster, most likely including inappropriate design, improper construction methods, and lack of control.

Figure 2.6 shows a detailed picture of the foundation of the apartment building, including part of the foundation piles. According to Gregory (2009), part of the explanation for the collapse is related to the design of the foundation. He refers to the chairman of the Japan Structural Consultants, who said that 'judging from Japan's standards, the posts of the building are too few and too thin. [...] in Japan where there are frequent earthquakes, the posts would be around 2 meters in diameter; but those of the collapsed building are about 50–60 centimeters; in addition, those posts are all hollow, where they should be solid'.

However, following the conclusion of an official investigation into the collapse of the building, the accident was determined to be due to improper construction activities related to the completion of an underground car park. Earth was excavated to make a 4.6-meter deep pit at one side of the building, and was being stored to heights of up to 10 meters on the other side of the structure. The weight of the excavated earth created a shear mode failure in the soil structure, damaging the foundation piles and causing them to fail. This situation may have been aggravated by several days of heavy rain.Probably the combination of inappropriate design of the foundation and improper excavation activities leading to an increased loading on the foundation piles caused the collapse to happen. As in many cases, here the problem could have been avoided by a better organised quality control during structural design and construction.

2.3 WRONG ESTIMATION OF LOADING

In the previously considered case of the Sleipner A platform, it was mentioned that the internal shear forces were underestimated during the design stage, although the external loading—the hydrostatic pressure—was correctly identified. In some cases, it might be difficult to correctly define the external loading on a structure. In some cases, inaccurate estimation of the loading might be due to the stochastic nature of external loading like wind pressure and earthquakes. In the case of new structures pushing the limits of earlier experience, like record span cable bridges or dazzling high skyscrapers, extrapolation of deemed-to-satisfy rules might not be sufficient as new and up-to-then unknown phenomenona might occur. In rare cases, wrong estimation of loading might simply have to do with erroneous static analysis. All of this will be illustrated by means of some case studies.

2.3.1 Roof collapses due to snow loading in Belgium, winter 2010-2011

Belgium has a moderate climate with rather cool summers and soft winters. For several decades, no significant snow precipitated during winters in Belgium. Due to this experience, snow loading was generally considered as a very low risk, and not much attention was paid to special measures like removal of melting water once the snow starts to melt.

Unfortunately, the 2010–2011 winter in Belgium was quite severe (according to local standards). In December 2010, 23 days of snow were reported, whereas the average is below five. By the end of 2010, a substantial amount of snow, in certain regions mounting to about 40 cm, caused several cases of collapsing roofs. The combination of reduced attention to snow loading during design and the exceptionally high snow loading in 2010 significantly contributed to the high number of collapses (around 40 in a few weeks time). A major aspect seems to be that designers seldom considered the so-called 'ponding effect' on flat roofs, while this is crucial information for a correct dimensioning of the emergency water evacuation system in case the normal water evacuation is blocked, e.g. due to ice formation as was the case here (Parmentier and Van de Sande 2011).

2.3.2 Tohoku earthquake and tsunami, Japan, March 11, 2011

On March 11, a magnitude 9.0 earthquake hit the east coast of Japan, also hitting the nuclear power plants in Fukushima. Although buildings and other structures could reasonably well withstand the impact of the earthquake (Japan has a long experience with earthquake resistant structures), it was rather the quickly following tsunami that caused most of the damage (and unfortunately thousands of casualties).

According to Clenfield et al. (2011), Japan has suffered 195 tsunamis since the year 400. Three in the past three decades had waves of more than 10 meters. A 7.6-magnitude quake in 1896 off the east coast of Japan even created waves as high as 38 meters, while an 8.6-magnitude quake in 1933 led to a surge as high as 29 meters.

The tsunami following the recent Tohoku earthquake in March 2011 also had wave heights up to 38 meters. Although Japan has invested the equivalent of billions of dollars on anti-tsunami seawalls which line at least 40% of its almost 35,000 km coastline and stand up to 12 meters high, the tsunami simply washed over the tops of some seawalls, collapsing some in the process (Onishi 2011): 'The height of seawalls varies according to the predictions of the highest waves in a region. Critics say that no matter how high the seawalls are raised, there will eventually be a higher wave. Indeed, the waves from Friday's tsunami far exceeded predictions for Japan's northern region'.

2.3.3 Collapse of the Tacoma Narrows Suspension Bridge, USA, 7 November 1940

All students studying structural behaviour will surely know about the Tacoma Narrows suspension bridge which collapsed in 1940. During the design of the Tacoma bridge however, the phenomenon later causing its collapse was simply not known to structural engineers. The bridge pushed the limits of the state-of-the-art, and, by failing, became a textbook example for avoiding similar failures in future.

After completion, the 1.6 km long Tacoma Narrows suspension bridge was the second largest in the world. On 7 November 1940 it collapsed (fortunately without any casualties) due to wind-induced vibrations, after only four months of service. The phenomenon causing the collapse is now known as resonance, and is caused by wind providing an external periodic frequency matching the natural structural frequency. Due to this accident, research efforts have been concentrated on bridge aerodynamics, influencing the designs of new long-span suspension or cable-stayed bridges.

2.3.4 Hyatt Regency walkway collapse, Kansas City, USA, 17 July 1981

On 17 July 1981, while hundreds of people were enjoying a music performance in the atrium of the Hyatt Regency, Kansas City, Kansas, USA, the suspended walkway suddenly collapsed, killing 114 and injuring many more (Petroski 1982, Natarajan 2007). The reason for the collapse was the failure of the connection between the upper deck and the suspension rods (Figure 2.7). While originally designed otherwise, the walkway was actually constructed with two separate rods (one for each deck, with the lower attached to the upper deck) instead of one single rod. As mentioned

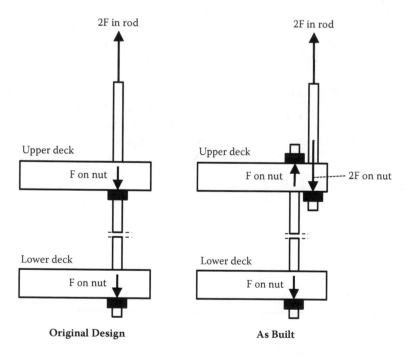

Figure 2.7 Hyatt Regency walkway: original design and situation as built.

by Natarajan (2007): 'After all the exotic forensic theories and scenarios had been exhausted, it turned out that the disaster was caused by an over-looked fact of simple high school level statics: That the substitution of two rods in place of one in the particular fashion it was done doubled the load at the nut' (Figure 2.7).

2.3.5 Thermal loading on thin marble façade cladding, Grande Arche, Paris

One of the many remarkable buildings in Paris, France, is the 110 meters tall Grande Arche, inaugurated in 1989. Less than 20 years after the inaugura-tion, serious problems were noticed with the façade cladding. Large thermal deformations were noticed, especially on the south side of the building, and the marble claddings became more brittle. The effect of thermal loading seemed to be clearly underestimated. Study revealed that during more than 150 days per year, the amplitude of the daily temperature variation on the south façade surpassed 40°C. Due to this significant repeated thermal load-ing, thin marble façade claddings show expansion bowing, loss of strength, and possibly detaching from the anchoring system (Grent et al. 2006).

2.4 INAPPROPRIATE ESTIMATION OF CREEP EFFECTS

When structures are loaded, structural deformations will occur. Normally, these deformations are very small, and can hardly be noticed by the naked eye. However, under sustained loading, deformations can increase in time. This is the so-called creep effect, which is very important in concrete struc-tures (Neville et al. 1983). The long-term deformation of concrete structures under sustained loading can typically be two to five times higher than the short-term deformation immediately after loading. Not duly considering the effect of the increasing deformation with time can lead to unwanted consequences.

In case of long-span bridges, the increased deflection due to creep effects can clearly become visible, especially when looking in longitudinal direc-tion. Creep effects can also be visible in other structures, as shown in Figure 2.8. Although these creep deformations will not directly impair the load bearing capacity of the structure, the excessive deformations are not esthetic and might give an unsafe feeling.

In apartment or office buildings, creep effects might lead to cracking in internal walls. These buildings are typically concrete skeleton buildings, with a load bearing structure consisting of columns, beams, and slabs. The internal walls are not bearing any load, and are typically made of a light material like gypsum blocks. These light walls cannot follow excessive

Figure 2.8 Long-term creep deformations of concrete parking structure.

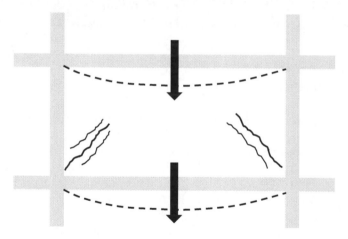

Figure 2.9 Cracking of light internal walls due to excessive creep deformations of concrete beams and slabs.

creep deformations of the beams or slabs on which they are standing, leading to possibly excessive cracking (Figure 2.9). Unfortunately, this damage pattern quite often occurs in apartment and office buildings. In many cases, it shows that the design did not duly consider creep effects, with estimated deformations following linear elastic theory only.

In tall buildings, unexpected situations can also occur when neglecting creep effects in concrete. Tall buildings typically consist of a central stiff core made in concrete, surrounded by a steel skeleton for the construction of the different floor levels. Due to creep and shrinkage of the concrete core,

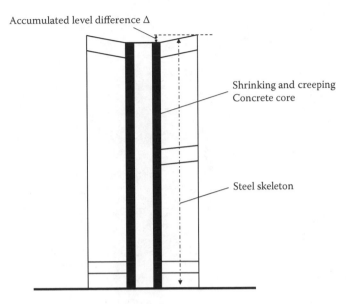

Figure 2.10 Effect of concrete core creep and shrinkage in tall buildings.

an accumulated level difference mounting up to several centimeters can occur at the highest floor levels (Figure 2.10). Although the stability of the tall building will not be in danger, slightly inclined floors can make it very uncomfortable to live or work at the highest floor levels.

2.5 INAPPROPRIATE MIX DESIGN

Design errors can also be made on the level of the concrete composition. An inappropriate selection of the required concrete type can be made, not duly considering the environmental aggressiveness. Or, in the case when the right concrete type has been selected, the detailed mix design can still be unsatisfactory. Both aspects will be explained in the following sections.

2.5.1 Inappropriate selection of concrete type

A good and durable concrete should be selected carefully considering the potential actions of the environment (e.g. frost, chlorides, carbonation, sulfates; see also Chapter 1). In most cases, the appropriate consideration of environmental actions and the subsequent selection of adequate concrete type is well-known and does not pose significant problems in concrete practice. However, there are situations where the appropriate selection of required concrete type in order to guarantee a durable structure is somewhat less obvious. Typical cases leading to repeated mistakes are swimming pools.

Figure 2.11 Degradation of concrete in the technical area underneath a swimming pool.

Quite often it is considered that no particular requirements have to be defined for the concrete type applied for the construction of swimming pools, reasoning that the concrete will be protected by a water-tight membrane protecting it from the chloride containing pool water. Apart from the significant impact of leaks in the membrane, this approach does not duly consider the fact that the main risk of concrete degradation in swimming pools is more typically situated in the technical areas underneath or alongside the pool itself (Figure 2.11). In these technical areas, a chloride containing environment is available, quite often at higher temperature (even up to 30°C). In many cases, water leaks containing chlorides are also occurring in pumps or pipes. These conditions can lead to an accelerated degradation of the concrete of the pool when an inappropriate concrete type (e.g. too high water/cement ratio) has been selected.

When studying the aggressiveness of the environment in order to select the appropriate concrete type, another shortcoming can be the underestimation of a combination of environmental actions. As an example, in a situation where carbonation can occur in combination with frost attack or sulfate attack, the combined effect can be much more severe than in the case of frost attack or sulfate attack alone. In special cases such as sewer water purification plants, the combined action of carbonation, frost attack, sulfate attack, and mechanical loading can lead to a significant acceleration of concrete degradation.

2.5.2 Bad mix design

Once the right concrete type has been carefully selected, things can also go wrong during the detailed mix design. Several problems can occur, for example:

- Inappropriate cement content, e.g. too high cement content in massive structures, leading to excessive heat of hydration and subsequent early age thermal cracking (see Chapter 4)
- Inappropriate workability of the concrete, leading to casting problems
- Inappropriate paste design, e.g. leading to excessive shrinkage or to segregation issues
- Inappropriate alkali content of the mix, possibly leading to detrimental processes (ASR, see Chapter 5)
- Incompatibility of different admixtures

A good mix design is important in order to successfully build in concrete. Due attention should be given to the mix design, including some experiments to verify the properties of the obtained material (initial type testing).

REFERENCES

Allen D.E. (1979) 'Errors in concrete structures', *Can. J. Civ. Eng.,* 6, 465–467.

Clenfield J., Humber Y. and Sato S. (2011) 'Tsunami Wall of Water Risk Known to Engineers, Regulators', Bloomberg, March 28, 2011.

Collins M.P., Vecchio F.J., Selby R.G. and Gupta P.R. (2000) 'Failure of an offshore platform', *Canadian Consulting Engineer*, March/April, 43–48.

Ioniță O.M., Țăranu N., Budescu M., Banu C., Romînu S and Băncilă R. (2009) 'Robustness of Civil Engineering Structures – A Modern Approach in Structural Design', *Intersections/Intersecții,* Vol.6, No.4, Article No.9, 99–114.

Gregory S., (2009), 'Rotten Foundations Cause Building Collapse in Shanghai', *The Epoch Times*, 1 July 2009.

Grent B., Schouenborg, B. and Malaga, K. (2006) 'Deterioration of thin marble cladding. A major international study', *Discovering Stone*, Issue 9, 22–28.

Kaminetzsky, D. (1991). *"Design and Construction Failures – lessons from forensic investigations"*, McGraw-Hill, New York.

Matousek M. and Schneider, J. (1976) 'Untersuchungen Zur Struktur des Zicherheitproblems bei Bauwerken', Institut für Baustatik und Konstruktion der ETH Zürich, Bericht No. 59, ETH.

Natarajan K. (2007) 'Forensic Engineering in Structural Design and Construction', CD Preprints of Structural Engineers World Congress 2007, Bangalore, India.

Neville A.M., Wilger W.H. and Brooks J.J. (1983) *'Creep of plain and structural concrete'*, Construction Press, London and New York, 1983.

Onishi M. (2011) 'Seawalls Offered Little Protection Against Tsunami's Crushing Waves', *New York Times*, March 13, 2011.

Parmentier B. and Van de Sande W. (2011) 'Collapsed roofs due to (exceptional) snow fall' (in Dutch), Het Ingenieursblad, 2, March-April, 55–56.

Petroski H. (1982) *'To Engineer is Human'*, St. Martin's Press.

Chapter 3

Errors during casting

3.1 INTRODUCTION

After a perfect design, things can go wrong during casting of a concrete structure. The concrete constituents can be badly proportioned or the mixing procedure can be inappropriate. By ignorance or by accident, aggressive substances could even be mixed into the concrete. On the construction site, reinforcement could be badly positioned or in insufficient quantities. The concrete could be badly vibrated or concrete vibration could even be omitted. On the level of the formworks, problems can occur due to insufficient strength or stiffness and due to leakage. When demoulding, surface damage could occur due to inadequate preparation of the formwork surface. After demoulding, curing measures could be insufficient. A more detailed overview of possible errors during casting is given in this chapter.

3.2 ERRORS DURING PROPORTIONING

A concrete plant has to take great care when proportioning the constituent materials following the recipe as determined during the mix design. The cement and water content, of course are of major importance, but also an accurate dosage of fine and coarse aggregates has to be guaranteed. Even more sensitive to variations, possibly with serious consequences, are the admixtures like accelerators and plasticizers. Typically, special concrete mixes containing sometimes carefully designed combinations of binder powders and admixtures, as is the case for self-compacting concrete, can be very sensitive to errors during proportioning. The generally increased sensibility of self-compacting concrete to mix variations is well-reported in previous documents (RILEM TC 188-CSC 2006, De Schutter et al. 2008).

A basic issue is to carefully control the water content of the concrete mix, taking into account the actual water content of the aggregates. In order to show the dramatic consequences when not doing so, consider a conventional concrete mix containing 350 kg cement and 175 kg water per cubic meter of concrete. Suppose that the concrete should contain 700 kg dry sand per cubic meter. Consider now the situation that the sand to be used is wet with a water content of 6%. When the concrete plant ignores the water content in the sand, 700 kg wet sand is proportioned per cubic meter of concrete. This means an actual proportioning of $700/1.06 = 660$ kg dry sand per cubic meter, while at the same time 40 kg extra water has been added as part of the wet sand. This brings the total water content to $175 + 40 = 215$ kg per cubic meter, and the water/cement ratio rises from 0.5 to $215/350 = 0.61$. As a consequence, different concrete properties will be obtained concerning workability, strength development, and durability.

The example is straightforward and most concrete plants will not make the error of ignoring water content in aggregates. However, related errors more often occur due to wrong determination of the water content (e.g. due to badly calibrated moisture sensors or due to measurements of non-representative samples) or due to deviations in the weighing equipment. Care should also be taken to avoid extra water addition within the truck mixer during transport after arrival at the concrete plant because the amount of added water cannot be accurately controlled. During transport of the freshly mixed concrete, care should also be taken to avoid rain water coming in. In case of sensitive mixes like self-compacting concrete, this extra (rain) water could lead to segregation of the mix, as experienced in at least one case known to the author.

Deviations in proportioned amounts of aggregates can also lead to modified concrete properties. In general, it can be said that proportioning errors of fine aggregates tend to have larger consequences than proportioning errors of the coarser aggregates. This can be illustrated by the information given in Table 3.1, showing the effect of variations in grading curves of aggregates on the obtained concrete properties (Taerwe 1996).

Due to the very small quantities in which they are proportioned, deviations in the dosage of admixtures like accelerators and plasticizers can have major effects. On the one hand, small absolute errors on their dosage easily

Table 3.1 Influence of variations in grading curves of aggregates on the obtained concrete properties

Particle size	Considered variation	Effect on concrete properties
15 mm	±20%	No significant effect
5 mm	±15%	Slight impact on workability of fresh concrete
2 mm	±10%	Modified workability of fresh concrete and strength of hardened concrete
0.2 mm	±10%	Important effect on fresh and hardened concrete properties

represent large relative errors. On the other hand, the effects can become significant due to possible side effects of the admixtures. This will be illustrated by two real cases.

3.2.1 Wrongly proportioned accelerator

Until recently, the application of calcium chloride was regularly used as an accelerator for concrete hardening in winter times. As a side effect, it is clear that the chlorides within the accelerator increased the risk of reinforcement corrosion. For this reason, nowadays chloride based admixtures are prohibited by the European Standard EN 206-1. In the 1990s, one specific case clearly illustrated these risks. In winter time in Belgium, the underground concrete basement walls of an office building were cast with the implementation of calcium chloride as an accelerator. Two months later, when the structure had climbed already a few levels, the basement walls started colouring brownish. Investigation showed that the chloride content in the concrete was significantly higher than what was generally accepted as a critical value with respect to reinforcement corrosion. It was shown that the applied dosage of the calcium chloride accelerator was significantly higher than the maximum allowed dosage in an attempt to keep up with the casting schedule in spite of the severe winter conditions. As a consequence, the reinforcement started corroding heavily, leading to the brownish discolouring of the walls. An attempt was made to extract the chloride by means of electrochemical techniques, however, without success. Finally, the partially completed concrete structure was torn down, and the contractor could start all over again.

3.2.2 Wrongly proportioned plasticizer

In another remarkable example, a water tower was built, consisting of a vertical concrete shaft on which an inverted cone reservoir had to be positioned. The shaft construction went on very smoothly, following a continuous operation with day and night shifts. However, after time, some horizontal rings became visible showing a sequence of lighter and darker concrete zones coinciding with the day and night shifts. Investigation showed that the darker zones showed lower compressive strength of the concrete below target values. Detailed investigation pointed to a wrong proportioning of a plasticizer. According to the instructions, for each concrete mix (produced on site) 30 cc of the plasticizer had to be added. The technician responsible for adding the plasticizer wrongly considered that 30 cc was about the volume of a beer glass leading to an over dosage of the plasticizer by almost a factor of ten! As a consequence, the hydration process of the concrete was totally disturbed, leading to insufficient strength. Also, in this case, only one solution was found: tear down the water tower, and start all over.

3.3 INAPPROPRIATE MIXING

The aim of mixing is to obtain a good and homogenous dispersion of all constituent materials within the concrete. The impact of mixing for mortar and concrete can be very significant and can affect the final properties (Williams et al. 1999, Chang and Peng 2001, Vickers et al. 2007). It should be kept in mind that concrete is a granular material, composed of a sometimes complicated combination of particles of different sizes ranging from micrometre to centimetre. Mixing sequences are very important, especially in case of mixes with high amounts of fine particles such as self-compacting concrete (Chopin et al. 2004, RILEM TC 188-CSC 2006). The adequate loading and mixing procedures are dependent on the mixer type, and should be determined on an individual basis, e.g. by some batching experiments (Jézéquel and Collin 2007).

When the mixing time is not long enough, especially the finer powder materials (cement and additions) will not be dispersed in a proper way and a good interaction between the fines and the admixtures might not be obtained. This can lead to insufficient workability or to insufficient strength development. On the other hand, when mixing time is too long, high air contents could be introduced into the mix, leading to a reduction of strength properties. In normal conditions, an optimal mixing time exists, depending on the type of concrete and the type of mixer.

Especially in the case of coloured concrete using pigment powders, a good mixing procedure should be determined in order to perfectly distribute the pigments into the material. Failing to do so can result in unacceptable colour variations on the final surface.

It is clear, however, that a completely uniform distribution of all particles and constituent materials of concrete can be achieved only in theory. Full homogeneity would require a perfectly uniform distribution of all components, not dependent on the sample size considered. It is commonly taken for granted that mixers produce homogenous concrete, because it is usually impracticable to test the entire batch (De Schutter et al. 2008). In reality, concrete mixers deliver fresh concrete of varying degrees of 'homogeneity' or 'uniformity' (Bartos and Cleland 1993), which is then practically impossible to improve during the concrete construction process.

3.4 AGGRESSIVE SUBSTANCES WITHIN THE MIX

While concrete durability is normally verified with respect to potential environmental actions, it is clear that degradation of concrete can also occur due to aggressive substances mixed into the concrete itself. In this case, the 'attack' comes from within the concrete, and not from the environment.

As the aggressive substances are within the concrete from the beginning, in some cases the degradation can go much faster than in cases where the aggressive substances have to enter the concrete from the environment. However, in many cases the degradation still depends on the availability or transport of water and/or oxygen. Some potential aggressive substances which have to be avoided in a concrete mix are listed hereafter, along with explanations of the kind of problems to be expected.

3.4.1 Sulfates

Sulfate attack in cementitious systems is a complicated issue and will be explained in more detail in Chapter 5. However, in short, the problem of sulfate attack can be summarized as follows. Portland cement consists of different clinker minerals, among which is the aluminate phase C_3A (tricalciumaluminate). As a result of a normal hydration process, mono-sulfate is formed within the cementitious system (see also Section 1.4.1 in Chapter 1). When this monosulfate is later enriched by sulfates, it will be transformed into ettringite (also called tri-sulfate). This conversion is accompanied by a volume increase because the ettringite is much richer in bound water. As a result, an internal pressure occurs within the hardened concrete, possibly leading to severe cracking (see Chapter 5 for more details on sulfate attack).

While the normal case of sulfate attack is caused by enrichment of the cement stone by sulfates coming from the environment, it could also happen that sulfates are mixed within the concrete during production. As a matter of fact, cement itself also contains sulfates. However, when cement conforms to actual code provisions, the risk of sulfate attack by the sulfates of the cement itself is practically excluded (except maybe the special situation of delayed ettringite formation in case of hydration at very high temperature, see further in Chapter 5).

Sulfates could also be added through additions such as ground granulated blast furnace slag or other fines like limestone powders. The sulfate content of these additions should also be limited, as typically required in relevant standards covering these materials.

A situation entailing more risk seems to be the application of recycled aggregates, e.g. in the form of masonry rubble. These materials, which typically originate from the demolition of houses, can contain sulfate impurities coming from the interior plasterwork (gypsum). Due attention should be given to avoiding these impurities because they will provide internal sources of sulfate within the concrete, easily leading to damage due to sulfate attack. Also in the case of aggregates coming from the sea (sea sand and sea gravel), due attention should be given to wash the aggregates with clear water in order to avoid contamination with sulfates and other aggressive substances.

It is good to recall in this section that the risk of sulfate attack can be significantly reduced by applying high sulfate resisting cement types, which contain a limited amount of aluminate phases. As a result, the hardened concrete will contain only minor quantities of monosulfate, so that even in the presence of sulfates, no significant conversion to ettringite can occur. However, some specific types of sulfate attack remain possible, even in this case.

3.4.2 Chlorides

As previously mentioned in the second paragraph of this chapter, a too high chloride content in concrete can initiate corrosion reinforcement. Further details on this corrosion process and on the critical chloride content to initiate corrosion will be given in Chapter 5. When producing a concrete mix, the total chloride content should be calculated based on the chloride content and the dosage of all constituent materials. When using sea aggregates, they should be washed with clear water in order to limit the chloride content. The chloride content of admixtures should also be carefully checked. In Europe, EN 206-1 even prohibits the use of chloride based admixtures.

3.4.3 Alkalis and potentially reactive aggregates

Some aggregates applied in concrete are potentially reactive when present in an alkali rich solution. In many cases, the aggregates contain potentially reactive amorphous silica, leading to the so-called alkali silica reaction (ASR). In Chapter 5, the expansive alkali silica reaction which leads to cracking and degradation of the concrete will be explained in more detail. The presence of potentially reactive aggregates alone is not sufficient for the occurrence of ASR. Water and alkali must also be sufficiently present for the expansive reaction to occur. As a consequence, the application of potentially reactive aggregates should also be combined with a careful control and limitation of the alkali content in the freshly mixed concrete.

Alkalis (Na_2O and K_2O, often combined in the equivalent alkali content Na_2O_{eq}, see Chapter 5) can be present in cement, admixtures, mixing water, and aggregates (especially in the case of sea aggregates). The application of special cement containing a very low alkali content (so-called LA cement or low alkali cement) can be required in case the use of potentially reactive aggregates cannot be avoided. However, it is a better approach to control the total alkali content per cubic meter of concrete. In this approach the application of an LA cement can be one element.

3.4.4 Free lime

Free lime, CaO, will hydrate when present in concrete and will form calcium hydroxide, $Ca(OH)_2$. This hydration process is expansive and can

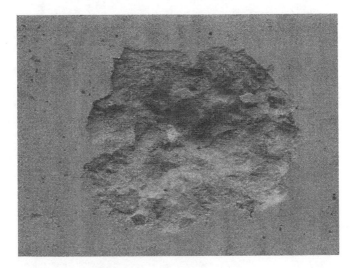

Figure 3.1 Pop-out due to free lime particles.

lead to cracking when the concrete cannot resist the generated stresses. For this reason, the free lime content in cement is well monitored and limited by provisions given in cement standards. As a consequence, contamination of the concrete by free lime present in cement rarely occurs in actual practice.

However, free lime can accidentally also be added to concrete due to a contamination of aggregates. In Belgium, this happened in May 2010, when a silo of limestone aggregates accidentally was contaminated with free lime particles (Moerman 2010, Marinus 2011). In total, about 100 tons of free lime were blended within 35,000 tons of limestone aggregate. The contaminated aggregates were delivered to large construction sites as well as to precast factories and concrete plants. In total, about 100 concrete structures have been affected by this contamination.

The free lime particles, which had a size ranging from 2 mm to 20 mm, expansively hydrated to calcium hydroxide after the addition of the mixing water. As this hydration process takes some time, at least part of the hydration took place after the concrete had already hardened. After some months, many customers started to complain about 'pop-outs' on the concrete surface, as shown in Figure 3.1. Although these pop-outs do not immediately impair the load bearing capacity of the structure, they at least provide an esthetical problem and possibly also increase the risk of accelerated durability issues due to the damage provided to the concrete cover.

3.4.5 Swelling aggregates

Similar to the case of free lime particles, aggregates in general can cause pop-outs on the concrete surface when some expansive reactions occur.

Figure 3.2 Disintegration of a concrete block containing non-stabilized slag particles.

A typical example is the case of aggregates containing pyrites. These minerals can undergo some expansive corrosion processes, leading to pop-outs as previously described in the case of free lime particles, however, now with some brown corrosion products at the location of the aggregate particle.

Care should also be taken when using by-products or waste-products as aggregates in concrete. It should be verified whether the alternative aggregate particles are stable enough so that expansive reactions and consequent cracking can be avoided. As an example, Figure 3.2 shows a disintegrating concrete block containing non-stabilized LD-slag as aggregate particles. LD-slag is an industrial by-product resulting from the transformation of hot metal into steel by oxygen refining. Most of the oxides formed are fixed into a basic slag built by the addition of lime and by the combustion of some iron and manganese. Consequently, LD-slag is a crystalline product containing mainly calcium silicate and combinations of lime and iron oxides. However, without any further treatment the LD-slag shows an insufficient volume stability, resulting in an expansive reaction and concrete damage. To overcome this problem, treated LD-slag has to be used. Methods to increase the volume stability are weathering with water or steam, thermal mixing with other products such as glass or sand, and cold mixing with incinerator bottom ashes or blast furnace sand (De Schutter et al. 2001).

3.5 WRONG PLACEMENT OF REINFORCEMENT

Accurately placing the designed amount of reinforcing steel is a key issue in concrete construction. It is advisable to carefully check the positioned steel cages before casting. As soon as the concrete is in place, it is nearly impossible to find out the real positioning and amount of the reinforcement. As illustrated hereafter, things can go wrong on different levels.

3.5.1 Wrong amount of reinforcement

It is straightforward that the designed amount of reinforcing steel has to be positioned, as outlined in the reinforcement plans developed during the design stage. Leaving out a few bars can have serious consequences such as excessive cracking and even structural failure. Due attention should be given to the anchorage and overlap zones in order to respect prescribed designs.

In case of steel fibre reinforced concrete, the prescribed amount of fibres has to be respected. During mixing, it has to be made sure that the right amount is added and that the fibres are evenly distributed within the whole batch. This brings us back to the issue of accurate proportioning and appropriate mixing (Sections 3.2 and 3.3 of this chapter). Nevertheless, statistical variations of the fibre content in the finalized structural element will always exist (Taerwe et al. 1997).

3.5.2 Wrong position of reinforcement

Steel reinforcing bars in structural concrete elements have to be positioned where they can efficiently take over tensile loading, leaving the compressive loading mainly to be taken by the concrete itself. In this way, a perfect symbiosis between steel and concrete is obtained, as has been shown in innumerable structures worldwide. Although the principle seems not too complicated, optimal solutions sometimes are not straightforward in the case of complex structures. This, however, is beyond the scope of this textbook.

In the following example, the consequences of a wrong positioning of the reinforcement will be illustrated. Figure 3.3 shows the structural concept of a balcony, constructed as a cantilever slab rigidly connected to the concrete skeleton of the building. When the balcony is loaded, it will bend downward. Tensile stresses will occur in the upper part of the balcony, compressive stresses in the lower part. Placing the reinforcing bars in a central position, as suggested in the figure, will thus not provide an adequate solution. This will not be an optimal position for carrying the tensile loads. The right solution in this case is to place the reinforcing bars towards the upper side of the cantilever slab leaving sufficient cover thickness.

The given example is a well-known case, to which only few errors will be made in practice. However, in less obvious cases these kinds of errors do occur. As an example, it is remarkable to see that foundation slabs (although

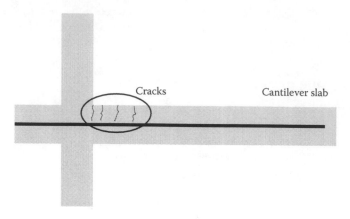

Figure 3.3 Inadequate positioning of reinforcement.

a rather obvious case) are sometimes diagnosed having an upside down positioning of the reinforcement cage. Most probably, this error is caused by the geometric analogy between foundation slabs and floor slabs. Only, in the case of foundation slabs the main loading is bottom-up (upward reaction forces of the underground), while it is top-down in case of floor slabs (live loading on the floor).

3.5.3 Too dense reinforcement

Sometimes, geometrical constraints are very strict and external loading relatively high. This can lead to high reinforcement ratios and, finally, to a very dense reinforcement cage as illustrated in Figure 3.4. Although from a mechanical point of view the situation might be structurally correct, the very dense reinforcement will make it very difficult or even impossible to correctly cast the concrete, filling the whole area between and underneath the reinforcement bars. This will lead to voids in the concrete element and to insufficient bond between concrete and steel, thus impairing the load bearing capacity of the element. Due to the voids and an increased porosity because the concrete cannot be duly compacted, the durability of the structural element will also be at risk, e.g. due to accelerated reinforcement corrosion.

Whenever dense reinforcement cannot be avoided, it is advised to find an adequate solution such as considering the application of self-compacting concrete. Self-compacting concrete has an ability to flow under its own weight, pass through or around a dense network of reinforcing bars, fill the required space or formwork completely, and produce a dense and adequately homogeneous material without a need for compaction (De Schutter et al. 2008).

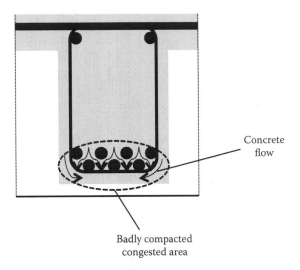

Concrete
flow

Badly compacted
congested area

Figure 3.4 Too dense reinforcement.

3.5.4 Insufficient cover thickness

In Chapter 1, the importance of respecting the cover thickness was explained in detail, showing that reducing the cover thickness by half can lead to a reduction of the service life (based on reinforcement corrosion) to 15% of the target service life. Thus, it is worthwhile to carefully control the real cover thickness provided during construction. Prescribed cover thickness values should be considered as minimum values and not as average values.

Due attention should then be given to a correct positioning of the reinforcement cage, avoiding shifting of the cage towards the formwork during casting as illustrated in Figure 3.5. This can be easily done by providing spacers with the right size and insufficient quantities and by inspecting over spacing before concrete placement. Failing to do so can lead to accelerated corrosion of the reinforcing bars which are too close to the surface, clearly making them visible due to expansive corrosion and subsequent concrete spalling as seen in Figure 3.6.

3.5.5 Wrong position of prestressing cables

In the case of prestressed concrete, the prestressing cables have to be accurately positioned. It should be kept in mind that the geometrical position of the cables, especially the curvature, is directly linked to the forces the prestressed cables will generate on the structural element. A detailed explanation of prestressed concrete is beyond the scope of this textbook; however, potential problems related to wrong positioning of a prestressing cable are illustrated in Figure 3.7 as an example.

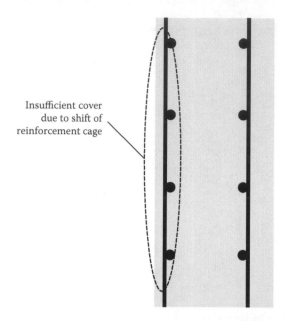

Insufficient cover
due to shift of
reinforcement cage

Figure 3.5 Insufficient cover thickness.

Figure 3.6 Reinforcement corrosion in case of insufficient cover thickness.

When the cable is positioned correctly, it will generate upward forces in the concrete beam, counteracting the downward forces provided by self-weight of the beam or by live load. This is how the prestressed cable increases the load bearing capacity of the concrete beam, compensating for (part of) the external loading, and making sure that no (or only limited) tensile stresses occur in the concrete.

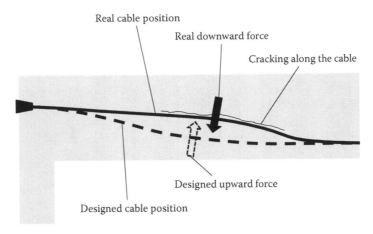

Figure 3.7 Wrong positioning of prestressing cables.

When the cable is not positioned correctly, now showing an inverted curvature, it will generate its forces in the opposite way. As a result, cracks can be easily generated in the concrete beam, potentially even pushing off the part of the beam underneath the cable. When positioning the cable ducts before casting the concrete, due attention should thus be given to avoiding displacements or uplifting during casting. Archimedes forces on the empty ducts in the liquid concrete should not be underestimated!

3.6 BAD COMPACTION AND OTHER PROBLEMS DURING CASTING

Traditional concrete should be compacted when placed into the formwork. As mentioned by De Schutter et al. (2008), 'It does not appear to be generally appreciated enough how great the distance is between compacted and uncompacted traditional low to medium consistence (workability) concrete'. Nevertheless, adequate compaction of fresh concrete is a fundamental requirement for good concrete construction (De Schutter et al. 2008). Failing to obtain adequate compaction of the fresh concrete is a major cause for insufficient performance of the final concrete product or element, including mechanical properties, durability behaviour, and esthetical aspects.

An appropriate level of compaction can be assessed by studying the volume of air voids in the hardened concrete, which in the case of adequate compaction normally will be around 1.5% to 2%. On construction sites, concrete is typically compacted by poker vibrators. In precast plants, other techniques can be used such as vibration tables or formwork vibrators.

It is very difficult to judge the efficiency of a compaction process in real time. The judgement typically is purely visual and based on the observation of air bubbles escaping from the surface of the concrete. Moreover, the compaction of fresh concrete on site is a complicated and laborious operation, leading to many problems. This is nicely illustrated by De Schutter et al. (2008): '*Compaction of fresh concrete of traditional consistence is often claimed to be the most unpleasant and tiring job on a typical construction site. In some countries it has become difficult to find workers prepared to carry this out. Supervision of the compaction process is also inherently difficult and it is therefore not surprising that a substantial proportion of concrete placed world-wide is, in reality, not adequately compacted*'. As a matter of fact, this conclusion was a driving force for the introduction of self-compacting concrete in the 1980s.

Defects caused by inappropriate compaction of the fresh concrete are often hidden, but can nevertheless significantly reduce the performance. In many cases, defects due to bad compaction become visible after striking of the formwork, as illustrated in Figure 3.8. Porous zones can be clearly defined and typically follow the position of reinforcing bars. *What normally follows is a repair of the affected concrete, usually called 'making-good'. [...] it has been recently estimated that around 30% of all UK concrete placed in-situ required some 'making-good'* (De Schutter et al. 2008). It has to be considered however that these repair actions ('making-good') are rather limited to obtaining a visually acceptable concrete surface, and do not upgrade the concrete to its initially intended performance.

Figure 3.8 also shows some casting joints, which can be partly linked to compaction problems, but also to a delay between different batches and inappropriate preparation of the surface of the previously already set or

Figure 3.8 Visual defects due to inappropriate compaction of a concrete wall.

hardened cast. This problem is also illustrated in Figure 3.9. When new concrete is placed on old concrete (already set or hardened), the old concrete should be free of dirt (sand, oil), dust and laitance (see also Section 3.8 of this chapter, bleeding). Failing to do so will result in a visible joint in the finalized concrete surface. More importantly, it will also lead to a weak connection between the old and new concrete.

During casting, unwanted inclusions can also be entrapped into the fresh concrete. In many cases, these inclusions remain invisible and unknown, although sometimes they become apparent on the final concrete surface, as illustrated in Figure 3.10 where some timber leftovers have been entrapped.

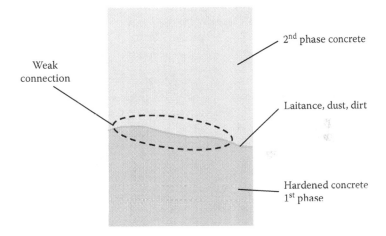

Figure 3.9 Casting joint.

Figure 3.10 Unwanted inclusions in concrete.

Another typical case is the inclusion of small steel leftovers (e.g. binding wire) which have fallen onto the formworks of slabs. When the formwork is not properly cleaned before casting the concrete, these steel parts will remain in the concrete near the surface, quickly leading to brown discolouring of the surface due to corrosion.

Mud and other dirt on the formwork, especially in case of slabs, can also lead to reduced concrete quality when it is intermixed with the concrete during casting, or to reduced bonding between different concrete batches or layers when the mud is enclosed in between. In winter, entrapment of ice can impair the final concrete quality.

3.7 PROBLEMS WITH FORMWORKS

3.7.1 Insufficient strength of formworks

Fresh concrete can exert considerable pressure on formwork, especially when the (vertical) casting speed is high. To a certain extent, pressures are close to hydrostatic pressures, levelling off later on due to setting effects. In the case of self-compacting concrete, which can be considered as a liquid and which enables very high casting speeds, pressures can be hydrostatic all the way up. Self-compacting concrete also enables new casting techniques, pumping the concrete under pressure into the formwork, bottom-up (De Schutter et al. 2008, Tichko et al. 2010). In that case, pressures can locally even exceed the hydrostatic pressures. In case the formworks have insufficient strength, serious accidents can happen during casting. As the safety of workers is then at risk, due attention should be given to safe formworks able to withstand the high casting pressures.

3.7.2 Too flexible formworks

More regularly occurring problems are related to too high deformability of formworks. When formworks are too flexible, displacement of the formwork panels can lead to leaks. More importantly, it can lead to incorrect geometry of the concrete element, and to concrete floors which seem to have very high deflections. Although this seldom leads to serious structural problems, from an esthetical or psychological point of view the resulting situation can be unacceptable.

3.7.3 Leaking formworks

When formworks are leaking, water and fine particles (cement, powders, fine sand) can be drained away, leaving unacceptable surface defects and honeycombed concrete as shown in Figure 3.11. A good tightness of the

Figure 3.11 Surface defects and honeycombed concrete due to leaking formworks.

formwork should be guaranteed, especially in the case of very flowable concretes like self-compacting concrete. As mentioned previously, too high flexibility of formworks can contribute to formwork leaking. Furthermore, incorrect use of poker vibrators can also cause movement of reinforcement and formwork and can contribute to leakage of paste.

3.7.4 Wrong positioning of formworks

Some details in the position of the formwork can have consequences for the mechanical and durability behaviour of the completed structure. Two examples with different consequences are shown hereafter.

The first example considers the connection between a concrete column and a rib floor as illustrated in Figure 3.12. As positioned here, the column formwork is somewhat too tall and penetrates the area of the concrete beam. As a result, the shear capacity of the beam near the supporting column will be significantly reduced, increasing the risk of shear failure.

The second example illustrates what can happen when concrete floor slabs are cast without making sure that the formwork shows a negative deflection while casting. After striking the formwork and loading the slab, the deflection will clearly increase. When the expected mechanical deflection of the slab in use is not (partly) compensated by a good positioning of the supporting formwork before casting the slab, the final deflection of the floor slab might visually become too high. In case of roof slabs, as illustrated in Figure 3.13, this might lead to stagnation of rain water, and increase the risk of ponding effects.

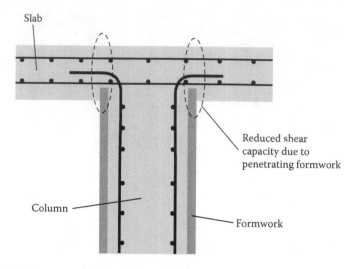

Figure 3.12 Wrong positioning of formwork at connection between column and rib floor.

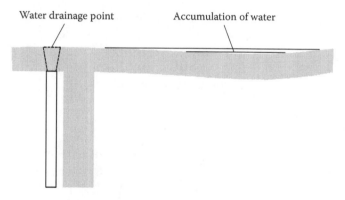

Figure 3.13 Wrong positioning of formwork (wrong inclination, not enough negative deflection during casting).

3.7.5 Demoulding problems

Problems can occur when striking the formworks is done too early. The concrete element should have enough strength to be self-supporting before formworks can be safely removed. As shown in Figure 3.14, premature demoulding of concrete floor slabs could lead to cracking near the beams. Even without cracking, the risk of higher deflections of the slab increases due to premature demoulding and thus mechanical loading of the slab. The earlier the slab is loaded, the higher creep effects will become.

The surface of formworks should also be well prepared before casting the concrete. This typically includes the application of demoulding oils, making

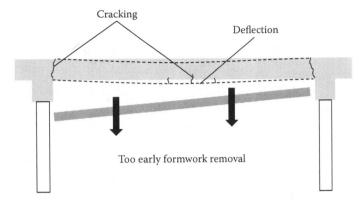

Figure 3.14 Possible consequences of premature demoulding in case of concrete floor slabs.

sure that the concrete will not stick to the formwork, making demoulding difficult and leading to possible damage of the concrete surface. Demoulding oils should be applied in appropriate quantities. Not enough demoulding oil will lead to a too high adherence between concrete and formwork material. Too much demoulding oil can lead to surface defects like air voids and discolouring.

3.8 DAMAGE IN PLASTIC STAGE

During the plastic stage, the first hours after casting, damage could be caused to the finished concrete surface by early loading, e.g. bikes driving too early on a very recently finished pathway, or even animals walking in the fresh concrete (see Figure 3.15). However, more frequently occurring problems in the plastic stage are plastic shrinkage, plastic settlement, and bleeding.

3.8.1 Plastic shrinkage

As explained in Section 1.4.6, freshly cast concrete has to be cured for a sufficiently long time period in order to maintain good conditions for the cement hydration (enough water), and in order to avoid early cracking due to premature water exchange with the environment. When the fresh concrete surface dries too fast due to insufficient curing, shrinkage can occur at this very early stage, possibly leading to cracking. At this stage, this phenomenon is referred to as plastic shrinkage and plastic-shrinkage cracking. Although it occurs most often in the dry summer, plastic shrinkage cracking can also happen in colder environments. While the risk is higher for concrete surfaces in outside environments, concrete floors inside a building or structure can also be exposed to plastic shrinkage, especially in environments with stronger wind or air movements.

Figure 3.15 Damage by animals walking in freshly cast concrete.

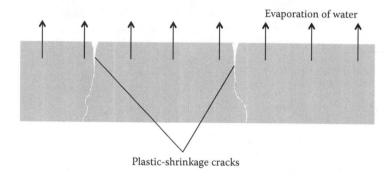

Figure 3.16 Plastic-shrinkage cracking in concrete slabs.

The most typical case in which shrinkage cracks occur is concrete slabs prior to final finishing and setting as illustrated in Figure 3.16. Due to moisture exchange with the environment, heavily accelerated by strong wind conditions, the exposed top surface of the concrete slab loses water. This drying process leads to a volume reduction. In totally free conditions, the plastic concrete could show a shortening even up to a few millimetres, which is an order of magnitude higher than the drying shrinkage occurring later after setting. This free plastic shrinkage cannot occur because it is restrained by the underlying concrete, which is not drying or at least at a much slower rate. As a result, plastic-shrinkage cracking can occur in the top zone of the concrete slab with crack openings which can go up to a few millimetres in severe cases. Although quite often plastic-shrinkage cracks are rather superficial (a few centimetres deep into the concrete), they can also run through the entire concrete slab depending on the boundary conditions.

A typical map-like plastic-shrinkage crack pattern is shown in Figure 3.17. However, in case of a strong wind in a quite steady direction, a pattern of mainly parallel plastic-shrinkage cracks can also occur as shown in Figure 3.18. In this case, as wind is a driving force, the plastic cracks are commonly referred to as wind cracks.

Figure 3.17 Typical plastic shrinkage map-like crack pattern in a concrete slab.

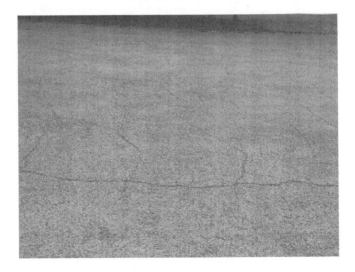

Figure 3.18 Parallel plastic-shrinkage cracks (wind cracks).

3.8.2 Plastic settlement and bleeding

Freshly cast concrete can show the tendency to bleed. Due to gravity, coarse aggregate particles can 'sink' somewhat downward as in a sedimentation process. This is called plastic settlement. As a consequence, water and fine particles (cement, powders) will go upward and a water film can be formed on the concrete surface, the so-called bleed water. The intensity of the plastic settlement and bleeding process is mainly dependent on the concrete composition (consistency and water content), but also on the height of the concrete element.

Bleeding occurs before setting and can counteract somewhat the effect of plastic shrinkage as long as the evaporation rate on the concrete surface is limited. However, bleeding leads to higher water content in the upper concrete, thus locally increasing the water/cement ratio and consequently porosity. Near the surface, a water rich paste layer is formed, called laitance. As a result, the concrete surface will be less resistant to frost, de-icing salts, and mechanical loading. After evaporation of the bleeding water, a dust rich surface is obtained. This dust can hinder the accurate operation of high tech machines such as automatic forklifts in storage buildings. The dust layer can also impair adherence of a new concrete layer cast in a later phase. Before casting new concrete on top of a previously cast hardened concrete, the laitance and dust have to be carefully removed.

In thicker concrete elements, plastic settlement and bleeding can lead to surface cracks following the upper reinforcement bars as illustrated in Figure 3.19. This is due to the fact that the reinforcement bars are fixed

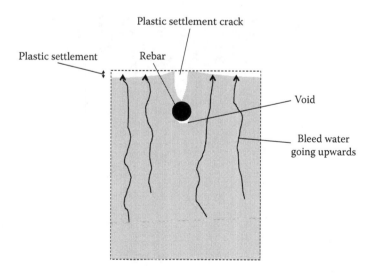

Figure 3.19 Plastic settlement and bleeding.

within a cage and cannot follow the settlement of the concrete. As a consequence, the concrete 'slides' off the reinforcement bars, leaving a crack opening on top of it. Underneath the reinforcement bars, a void can be created. This phenomenon leads to mechanical problems due to the reduced bond between the steel reinforcement and the concrete, as well as to durability problems due to accelerated ingress of aggressive substances with almost immediate and open access to the reinforcement bars.

While bleeding on top surfaces are a regularly occurring and well-known phenomenon, water movements can also occur between the concrete and the formwork. This can lead to small stripes, also called 'sand stripes', resulting from the upward water movement carrying small powder and sand particles. The risk of sand stripes increases in the case of a too high water content in the concrete and in the case of too long vibration.

REFERENCES

Bartos P.J.M. and Cleland D.J. (Eds.) (1993) 'Special Concrete: Workability and Mixing', London, E&FN Spon.

Chang P.-K. and Peng Y.-N. (2001) 'Influence of mixing techniques on properties of high performance concrete', *Cement and Concrete Research*, 31, 87–95.

Chopin D., de Larrard F. and Cazacliu B. (2004) 'Why do HPC and SCC require a longer mixing time?', *Cement and Concrete Research*, 34, 2237–2243.

De Schutter G., Bartos P., Domone P. and Gibbs J. (2008) 'Self-Compacting Concrete', Whittles Publishing, Caithness, UK, CRC Press, Taylor & Francis Group, Boca Raton, USA, ISBN 978-1904445-30 2, USA ISBN 978-1-4200-6833-7, pp. 296.

De Schutter G., Audenaert K., Sichien J. and De Rouck J. (2001) 'Experimental study of the incorporation of LD-slag in massive concrete armour units for breakwaters', 2nd Int. Conf. on Engineering Materials (Eds. Nagataki et al.), San Jose, 101–112.

Jézéquel P.-H. and Collin V. (2007) 'Mixing of concrete or mortars: Dispersive aspects', *Cement and Concrete Research*, 37, 1321–1333.

Marinus J. (2011) 'Pop-outs due to aggregates contaminated with free lime. Scientific study of the risks' (in Dutch and French), Beton, 209, January, 32–37.

Moerman B. (2010) 'Contaminated concrete threatens construction sites' (in Dutch), De Standaard, 18 September 2010, 2.

RILEM TC 188-CSC (2006) 'Casting of self-compacting concrete', Final report of RILEM Technical Committee 188-CSC, RILEM Report 35, RILEM Publications sarl.

Taerwe L. (1996) 'Concrete Technology' (in Dutch), Lecture notes, Ghent University, Belgium.

Taerwe L., Van Gysel A., De Schutter G. and Vyncke J. (1997) 'Experimental determination of the steel fibre content in fresh and hardened concrete used for concrete slabs on grade', Proceedings of the Asia-Pacific Conference on Fibre Reinforced Concrete, Singapore, August, 255–261.

Tichko S., Van De Maele J., Vanmassenhove N., De Schutter G., Vierendeels J., Verhoeven R. and Troch P. (2010) 'Numerical modelling of the filling of form-works with self-compacting concrete', in *Advances in Fluid Mechanics* VIII, Eds. Rahman and Brebbia, WIT Transactions on Engineering Sciences, Vol. 69, WIT Press, ISBN: 978-1-84564-476-5, 157–168.

Vickers T.M., Farrington S.A., Bury J.R. and Brower L.E. (2005) 'Influence of dispersant structure and mixing speed on concrete slump retention', *Cement and Concrete Research*, 35, 1882–1890.

Williams D.A., Saak A.W. and Jennings H.M. (1999) 'The influence of mixing on the rheology of fresh cement paste', *Cement and Concrete Research*, 29, 1491–1496.

Chapter 4

Actions during hardening

4.1 INTRODUCTION

One of the major difficulties with concrete as a construction material is the volume instability as a function of time. Due to several phenomena, often occurring simultaneously, the dimensions of a concrete element evolve in time even without external mechanical loading. When the resulting deformations are restrained, cracking and damage can occur in the structural concrete element. Especially during the hardening process, the cracking risk can be quite high, because in this stage the young concrete already has a relatively high stiffness and still a relatively low strength (see also Chapter 1).

As a consequence of the hydration process, water is chemically bound. The hydration products have a smaller volume than the reacting products, which is called chemical shrinkage. Furthermore, the consumption of water leads to some internal drying, especially in the case of binders with a low water/cement ratio. Consequently, a volume reduction is noticed, called autogenous shrinkage. Further details are given in Section 4.2 of this chapter.

As soon as the hardening concrete element is exposed to external drying, water is typically exchanged from the material to the environment, as the environment normally is dryer than the hardening concrete. The loss of water also gives rise to a volume reduction of the concrete, called drying shrinkage. In wet environment, the concrete element can show expansion instead of shrinkage. Further details are given in Section 3 of this chapter. In addition to drying shrinkage, concrete also shows carbonation shrinkage (Neville and Brooks 2010). Carbonation shrinkage can lead to damage in autoclaved aerated concrete as explained in Section 5.2.3 of Chapter 5.

Another effect of the hydration process is the exothermal production of hydration heat. This heat energy will cause a temperature increase of the concrete element. In a typical situation, the core of the element will show a higher temperature than the surface of the element, leading to thermal gradients and thermal stresses. As soon as the hydration process slows, the concrete element will start cooling down, causing thermal shrinkage. The phenomenon of early-age thermal cracking during hydration will be further discussed in Section 4.4 of this chapter.

It should be well considered that hardening concrete can be very vulnerable to crack formation. During hardening, stresses are quite often the result of an imposed deformation, e.g. due to shrinkage or thermal effects. When the imposed deformation is restrained, the resulting stresses are proportional to the level of restrained deformation and to the stiffness of the concrete (Young's modulus). When the concrete tensile strength is not sufficient to withstand the tensile stress, cracks will occur. At early age, the Young's modulus of concrete develops relatively much faster than the strength (De Schutter and Taerwe 1996). As a consequence, during hardening, relatively large stresses can occur due to restraint of imposed deformations while the concrete still has a relatively low strength. This fundamentally explains why hardening concrete is so sensitive to early-age crack formation.

4.2 AUTOGENOUS SHRINKAGE

4.2.1 Mechanism

During the hydration process of cement, the reaction between cement clinker minerals and water leads to the formation of hydration products, as also explained in Chapter 1. These chemical transformations are accompanied by a volume reduction, as the resulting volume of the reaction products formed is smaller than the sum of the volumes of the reacting cement and water, as illustrated in Figure 4.1. A volume reduction is thus obtained as a result of the hydration process. This volume

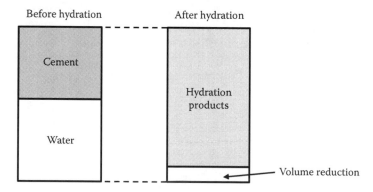

Figure 4.1 Chemical shrinkage.

reduction is called the chemical shrinkage (Koenders 1997, Bjøntegaard 1999, Tazawa 1999).

Due to the chemical shrinkage, the pore volume inside the hardening cement paste will increase with increasing degree of hydration. When no external water supply from the surrounding environment is provided, the water content in the pore volume decreases, while the air content increases. This is called self-desiccation. As a result of chemical shrinkage and self-desiccation, the air pressure in the pores will decrease, and will equilibrate with the capillary tension. The increasing capillary tension will cause a macroscopic volume reduction of the hardening cementitious material, which is referred to as autogenous shrinkage (Koenders 1997, Bjøntegaard 1999, Tazawa 1999). In a more complete picture, it has to be mentioned that autogenous shrinkage is influenced by several mechanisms: changes in the surface tension of the solid gel particles, disjoining pressure, and capillary tension (Kovler and Jensen 2007). Autogenous shrinkage does not include any effects on the volume caused by external mechanical loading, temperature variations, or loss or ingress of substances. Although autogenous shrinkage can be expressed as a percentage of volume reduction, a more common way to quantify autogenous shrinkage is to express it as a one-dimensional length change, the autogenous shrinkage strain.

Quite often, autogenous shrinkage is only considered after initial setting, excluding volume changes generated in the fresh state of the cementitious material. The reasoning behind this limitation is the fact that autogenous shrinkage is typically considered for the study of early-age cracking. Shrinkage stresses are only initiated once the material shows a certain stiffness, which begins after percolation of the microstructure. As this percolation coincides with the phenomenon of setting, the time of initial setting is typically specified as the starting point of autogenous shrinkage. In a more general way, this starting point is referred to as 'time zero' (Miao et al. 2007). In order to determine the value of time zero, estimations can be made based on the Vicat needle test, compressive strength results, heat of hydration, or ultrasonic transmission measurements (Bentur 2003, De Schutter 1996, De Schutter and Taerwe 1996, Robeyst 2008). For a detailed discussion on this topic, reference is made to scientific literature.

The correspondence between chemical shrinkage and autogenous shrinkage is illustrated in Figure 4.2. Immediately after water addition, the hydration process evolves, introducing chemical shrinkage into the cement-based system. As soon as setting initiates, the effects of chemical shrinkage and self-desiccation cause a macroscopic volume reduction of the hardening material, called autogenous shrinkage, which is much smaller than the purely chemical shrinkage. The difference between chemical shrinkage and autogenous shrinkage represents the volume of voids formed into the system due to hydration.

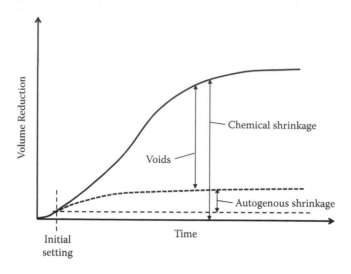

Figure 4.2 Correspondence between autogenous and chemical shrinkage.

Autogenous swelling peaks can also be noticed at very early age (Baroghel-Bouny et al. 2006). Different explanations can be given for the occurrence of swelling peaks, including the effect of mineral additions such as limestone filler on the disjoining pressure (Esping 2008, Craeye et al. 2010). The fineness of the filler is an important factor for this swelling behaviour, although the nature of the filler also seems to have an influence. Super plasticizers also interact with the fillers, influencing the swelling behaviour.

When the concrete element can freely deform without any restraint, autogenous shrinkage (and shrinkage in general) will not cause any stresses and thus no cracking or any other damage. However, as concrete elements are almost always connected to other structural elements or foundations leading to so-called external restraint, autogenous shrinkage will give rise to early-age stresses and possibly cracking.

Even when the hardening concrete element is not connected to any other restraining structural element, internal restraint could also occur due to differences in hydration rate. Due to a higher temperature caused by the heat of hydration, the core of a concrete element can hydrate faster than the surface zone, leading to a differential development of the autogenous shrinkage. As the core hydrates faster, autogenous shrinkage is also developing faster. The shrinkage of the core, however, is retrained by the surface zone, which is shrinking slower. In this case, the resultant effect on stress formation and cracking has to be studied in combination with the effect of the heat of hydration (see Section 4.4 on thermal shrinkage). In the case of massive structures, however, it can be said that the concrete normally does not have a very low water/cement ratio, so the effect of autogenous shrinkage could be neglected in comparison with the thermal effects.

4.2.2 Influencing parameters

Autogenous shrinkage strains of concrete can widely range from about 50 mm/m to about 500 mm/m or even larger. The main influencing factor seems to be the water/cement ratio of the concrete, although other influencing factors can be mentioned.

4.2.2.1 Mineral composition of cement

As the chemical shrinkage is a driving force for the occurrence of autogenous shrinkage, it is clear that the mineral composition of the cement is an important influencing factor. It is also clear that chemical and autogenous shrinkage will depend on the state of the hydration process with increasing values for higher degrees of hydration. Tazawa (1999) clearly showed that C_3A and C_4AF have a substantial influence on autogenous shrinkage.

4.2.2.2 Mineral additions and chemical admixtures

Some mineral additions such as silica fume can increase the autogenous shrinkage, while other mineral additions such as fly ash can lead to a reduction (Tazawa 1999). The effect is dependent on the nature and the fineness of the mineral addition possibly leading to swelling peaks and can be further influenced by the presence of super plasticizers (Craeye et al. 2010).

4.2.2.3 Water/cement ratio

Autogenous shrinkage increases with decreasing water/cement ratio. The effect of chemical shrinkage and self-desiccation is more pronounced in cementitious systems with a low water content relative to the cement content. This is totally different from drying shrinkage, which shows increasing shrinkage values for increasing water/cement ratios (see Section 4.3). As a result, autogenous shrinkage is a more important issue for structural elements made with high strength concrete, typically showing a low water/cement ratio (0.35 or lower) and incorporation of silica fume, while it is often neglected or implicitly incorporated in drying shrinkage models for normal strength concrete with higher water/cement ratios (0.45 and higher).

4.2.2.4 Paste volume

Shrinkage occurs in the cement matrix, while the aggregates in mortar and concrete do not show any shrinkage behaviour. As a result, as is the case for drying shrinkage as well, autogenous shrinkage increases with increasing paste volume in the concrete.

4.2.3 Mitigation

The most straightforward measure to reduce autogenous shrinkage is to increase the water/cement ratio. This of course is not always possible, especially in cases where a (very) high concrete strength is required. Furthermore, in high strength concrete, quite often silica fume is applied in order to refine the pore structure and further increase the strength. Unfortunately, this also further increases the autogenous shrinkage. Some reduction in autogenous shrinkage could be obtained by a good selection of the cement type, avoiding a high C_3A content, and by a well considered combination of mineral additions and super plasticizers.

In case of high strength concrete, special 'internal curing' measures might be needed (Kovler and Jensen 2007). Internal curing is defined as the incorporation of a component, which serves as curing agent, to the concrete mixture. Internal curing can be classified into two categories: internal water curing (or water entrainment) and internal sealing. Internal water curing seems to be the most effective in order to mitigate autogenous shrinkage. This method consists of the incorporation into the cementitious system of a curing agent serving as an internal water reservoir. The water is gradually released as the concrete dries out due to self-desiccation. Super absorbent polymers (SAP) and water-saturated fine lightweight aggregates (LWA) are mostly considered as internal water reservoirs.

The effect of internal curing (or any other method to reduce or mitigate autogenous shrinkage) can be measured by means of specially developed test methods. Autogenous shrinkage can be experimentally studied by means of different methods, including volumetric and linear measurements (Tazawa 1999, Loukili et al. 2000, Kovler and Jensen 2007).

4.2.4 Example

As an example, the application of high performance concrete for a bridge deck is considered, as discussed by Craeye et al. (2011). The effect of the addition of SAP on the autogenous shrinkage of the high performance concrete, as measured by means of a vertical dilatometer for concrete, is shown in Figure 4.3 (Craeye et al. 2011). A high performance reference concrete (REF) was considered with the composition given in Table 4.1. The water/cement ratio is 0.32. The water/binder ratio is 0.30. The super plasticizer

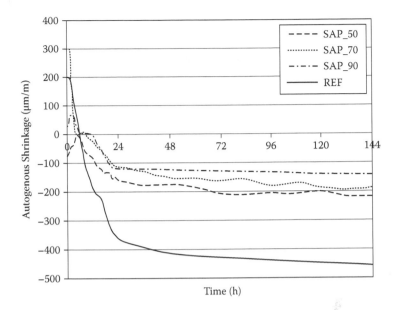

Figure 4.3 Reduction of autogenous shrinkage by means of SAP (Craeye et al. 2011).

Table 4.1 Concrete composition (kg/m³)

Component	REF	SAP50	SAP70	SAP90
Cement CEM I 52.5 R	475	475	475	475
Sand 0/5	588	461	410	359
Porphyry 2/6	482	482	482	482
Porphyry 6/20	677	677	677	677
Silica Fume	25	25	25	25
Super plasticizer	3.125	1.875	1.333	0.667
Water	150	150	150	150
Internal curing water	-	50	70	90
SAP	-	1.11	1.56	2.00

is a polycarboxylic ether. Porphyry coarse aggregates have been used, with particle size between 2 mm and 20 mm. Some silica fume was added.

This reference concrete, with a low water/binder ratio and containing silica fume, shows a high autogenous shrinkage strain, as shown in Figure 4.3, with a value up to 450 mm/m after 144 hours. Due to this high autogenous shrinkage in combination with the effects of the heat of hydration, early-age cracking can be expected when this concrete is applied for a bridge deck (Craeye et al. 2011).

In order to mitigate the autogenous shrinkage and to prevent early-age cracking, internal curing can be applied by means of SAP. The applied SAP has a water absorption capacity of 45 g/g after 5 minutes (the approximate mixing time). Based on this absorption level, the required SAP content is estimated, aiming for an amount of internal curing water equal to 50 kg/m^3 (SAP50), 70 kg/m^3 (SAP70), and 90 kg/m^3 (SAP90). This leads to a corresponding SAP amount of respectively 1.11 kg/m^3, 1.56 kg/m^3 and 2.00 kg/m^3, as shown in Table 4.1. The extra (internal) curing water has to be added to the concrete during mixing. In order to obtain 1 m^3 of concrete, the sand content is reduced accordingly.

From Figure 4.3, it is clear that a significant reduction of the autogenous shrinkage is obtained, although full mitigation is not reached. In parallel however, other aspects have to be studied, as the addition of SAP and internal curing water will also influence mechanical and thermal properties. More details can be found in Craeye et al. (2011). As a conclusion, however, it is found that due to the internal curing and its significant effect on autogenous shrinkage strains, early age cracking within the high performance bridge deck can be prevented.

4.3 DRYING SHRINKAGE

4.3.1 Mechanism

Cement paste typically contains different types of water: chemically bound, physically adsorbed, and free water (Hansen 1986). The chemically bound water has become part of the hydration products, as a result of the chemical reactions between cement and water. Physically adsorbed water can be found in the gel pores, adsorbed at the surface of the hydration products. Free water is present in the capillary pores. Physically adsorbed water and free water are also referred to as evaporable water, which can be removed from the cement paste by drying. The amount of evaporable water is strongly dependent on the water/cement ratio and the degree of hydration.

Drying shrinkage of concrete occurs when evaporable water is removed to a non-saturated environment. First, the free capillary water is removed, leading to a volume reduction of the concrete. When the capillary pores are empty, the physically bound water will also be slowly expelled, significantly increasing the volume reduction. Although from a phenomenological point of view the origin of drying shrinkage is straightforward (loss of water to a non-saturated environment), the contributing fundamental mechanisms at nano scale are not fully clear. Different mechanisms seem to contribute: capillary stress, disjoining pressure, movement of interlayer water, and changes in free surface energy (Mindess and Young 1981, Hansen 1987, Bisschop 2002).

Although drying shrinkage can lead to a volume reduction, drying shrinkage is most commonly expressed as a linear drying shrinkage strain. For concrete, drying shrinkage strain values typically ranging between 200 mm/m and 600 mm/m can be obtained, depending on several parameters.

As the concrete element loses water at the exposed surfaces, the drying process is controlled by a diffusion process, requiring water to diffuse from the inner parts of the concrete element to the exposed surfaces. This diffusion process is quite slow so that drying of concrete is a time-consuming process. Furthermore, the drying process of concrete elements is not uniform, inducing internal moisture gradients, and leading to higher shrinkage near the exposed surfaces and lower shrinkage in the core. This non-uniform shrinkage leads to internal stresses due to internal restraint, possibly leading to shrinkage cracks. The degree of self-restraint depends on specimen size (Wittmann 2001). In most cases, however, drying shrinkage cracking due to internal restraint is very limited. Internal microcracking can occur, however, due to shrinkage restraint by non-shrinking aggregate particles (Bisschop 2002). For concrete structures, external restraint of drying shrinkage deformations are more important, and can typically lead to significant cracking, e.g. in concrete floors and linear concrete elements (see examples in Section 4.3.4.).

Rewetting of (partially) dried concrete elements will lead to increase in volume or swelling. However, part of the drying shrinkage will not be recovered, which is called irreversible shrinkage (Figure 4.4). This probably is related to rearrangements in the CSH structure. When physically adsorbed water is expelled, the distance between hydration products is somewhat reduced. As a result, additional chemical connections can be formed, which prevent full shrinkage recovery. The irreversible drying

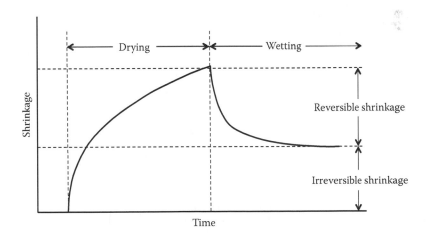

Figure 4.4 Reversible and irreversible drying shrinkage.

shrinkage can amount to up to 60% of the drying shrinkage, depending on the degree of hydration at the onset of drying. Carbonation during drying can also increase the irreversible drying shrinkage.

4.3.2 Influencing parameters

4.3.2.1 Parameters related to the concrete composition

Drying shrinkage of concrete is largely dependent on the paste volume, or alternatively, on the aggregate content. As the paste is the actively shrinking component, a reduction in paste content will reduce the overall shrinkage values. Furthermore, drying shrinkage of the cement paste is partly restrained by the non-shrinking aggregates (Bisschop 2002). Based on experiments, Powers (1949) found the following relation between concrete shrinkage ε_{cs} and cement paste shrinkage ε_{cps}:

$$\varepsilon_{cs} = \varepsilon_{cps} \left(1 - V_a\right)^n \tag{4.1}$$

with V_a the volume fraction of the aggregates and n an exponent between 1.2 and 1.7. When the volume fraction of the aggregates is 0.7, and considering $n = 1.7$, it is found that the concrete shrinkage is about 13% of the paste shrinkage. However, as already mentioned, stress concentrations will occur around the aggregate particles due to internal restraint.

The larger the water/cement ratio, the larger the drying shrinkage of the concrete will be. For a higher water/cement ratio, more free water will remain in the concrete after hydration, filling a more extended pore system. As a consequence, more water can be exchanged with the environment, increasing the drying shrinkage. In many codes, the effect of the water/cement ratio is implicitly implemented by considering the concrete strength.

The cement content affects the drying shrinkage by its influence on the paste volume or on the water/cement ratio. The cement type only seems to have a small effect on the ultimate drying shrinkage strain (Mindess and Young 1981, Bisschop 2002), although the time development of drying shrinkage strain can be influenced by the reaction rate of the cement. The addition of puzzolans (silica fume) can also influence the drying shrinkage, although part of this influence might be ascribed to autogenous shrinkage due to long-term puzzolanic activity, which is difficult to separate from the drying shrinkage measurements. Drying shrinkage also can vary with the application of super plasticizers, depending on type and amount. The composition of the mixing water also influences shrinkage properties: a lower alkali content leads to smaller shrinkage values (Beltzung et al. 2001).

4.3.2.2 Geometrical parameters

The time development of drying shrinkage is significantly influenced by the geometry of the element, due to the key role of the exposed surfaces. This influence is often taken into account by considering a geometrical parameter such as the fictitious thickness h_f:

$$h_f = \frac{A_c}{u/2} \tag{4.2}$$

with A_c representing the cross section area of the element and u the circumference. The smaller the fictitious thickness, the faster the drying shrinkage will develop. However, the fictitious thickness will only show a minor influence on the ultimate drying shrinkage strain at very long term (Almudaiheem and Hansen 1987, Bissonnette et al. 1999, Bisschop 2001). Concerning the effect of specimen size, some disagreement exists in literature. For smaller concrete specimens showing the same geometry, a somewhat smaller ultimate drying shrinkage value has been reported (Hobbs and Mears 1971). However, Mindess and Young (1981) theoretically reason that larger specimens will dry slower and the average degree of hydration that will be reached will be higher, leading to a somewhat reduced shrinkage value. Almudaheem and Hansen (1987) and Bissonnette et al. (1999) did not find any size effect in case of paste and mortar.

As explained before, shrinkage stresses will occur in the drying concrete element due to the moisture gradients and the self-restraint of the different shrinkage levels in the core and near the surface. For concrete practice, however, the drying shrinkage strain is mostly considered to be homogenous across the element section and shrinkage stresses will be caused due to external restraint.

4.3.2.3 Atmospheric parameters

For a lower relative humidity of the environment, higher ultimate drying shrinkage values will be obtained. As an example, for a typical concrete with a water/cement ratio about 0.5, an environmental relative humidity of 50% will lead to an ultimate drying shrinkage strain of about 500 to 600 mm/m, while in the case of 80% relative humidity, it will be about 300 mm/m.

The influence of drying temperature on the ultimate drying shrinkage strain is not very significant, although a lower drying temperature might lead to a somewhat smaller shrinkage strain (Bisschop 2001).

4.3.2.4 Technological parameters during execution

When a concrete element is exposed to drying starting from an earlier age, the ultimate drying shrinkage value will be higher (Therrien et al. 2000).

For a lower degree of hydration, the pore structure will be coarse and more evaporable water will still be available, leading to higher shrinkage values. The duration of the curing period thus is important for the drying shrinkage strains occurring as soon as curing is ended. This is not always well implemented in design codes.

It can also be mentioned that a lower curing temperature seems to lead to larger ultimate drying shrinkage strain (Bisschop 2001). This probably is related to the effect of temperature on the hydration process and thus on the value of the degree of hydration at the end of the curing period.

4.3.3 Mitigation

Considering the influencing parameters as mentioned in the previous section, some measures can be listed in order to reduce drying shrinkage strains in concrete elements:

- In order to reduce drying shrinkage strains, the water/cement ratio of the concrete has to be limited. However, as explained in the section on autogenous shrinkage, a strong reduction of the water/cement ratio will significantly increase autogenous shrinkage values.
- The paste volume of the concrete should be limited, which means that the aggregate content should be increased. This should be considered in view of other requirements, such as workability. In case of self-compacting concrete, a higher paste volume and lower aggregate content is quite common, influencing shrinkage behaviour (De Schutter et al. 2008). This does not mean however that self-compacting concrete always shows higher shrinkage strains.
- Curing of the fresh concrete should be carefully done and maintained long enough. Insufficient curing will lead to a lower degree of hydration and to higher drying shrinkage values.
- A good casting scheme can also contribute to a lower risk of drying shrinkage cracking. As shrinkage stresses are mainly caused by external restraint of shrinkage deformation, an optimized casting sequence, reducing external restraint, can be very helpful. Constructive measures like joints, reducing the length of linear elements, or the size of individual floor slabs, have to be duly investigated.

4.3.4 Example

Drying shrinkage cracking quite often occurs in concrete slabs on grade (often in combination with plastic shrinkage cracking). The size of the individual slabs is important, with larger sizes leading to increased cracking risk. Joints have to be provided or saw cuts have to be made shortly after

Figure 4.5 Drying shrinkage cracking in walkway.

casting and setting. In the example of Figure 4.5, joints have been provided at regular short distances in the longitudinal direction of the concrete walk way. However, in the transversal direction, no joints or saw cuts have been provided, yielding slab elements which are too long to avoid drying shrinkage cracking due to external restraint. As a result, a major crack occurred about halfway, running along the walkway.

Drying shrinkage cracking also typically occurs in linear elements such as New Jersey concrete barriers employed to separate lanes of traffic, as illustrated in Figure 4.6. Heavy reinforcement in these barriers will not prevent cracking due to drying shrinkage, but can help distribute cracks and limit crack widths. This principle is also followed in the case of continuously reinforced concrete highways without joints: cracks are not avoided, but rather controlled due to heavy reinforcement.

It is important to mention that the occurrence of shrinkage cracking is influenced by the relaxation behaviour of the concrete (which is related to the creep behaviour). Consider the theoretical example of a shrinking concrete element as illustrated in Figure 4.7, completely restrained at its ends. As a result of the restrained shrinkage deformation, tensile stresses will be generated in the concrete element. However, due to the relaxation properties of the concrete, the tensile stresses will be reduced. When estimating the effect of shrinkage on stresses in concrete structures, creep and relaxation properties should be duly considered, as purely elastic calculations might lead to wrong conclusions (overestimation of stresses in the example of Figure 4.7, but possibly underestimation in other cases).

Figure 4.6 Drying shrinkage cracking in New Jersey concrete barriers.

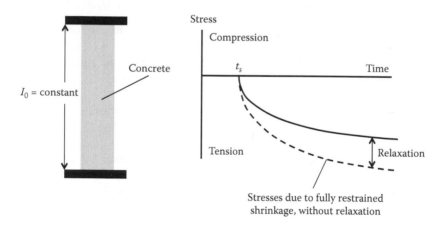

Figure 4.7 Influence of relaxation on shrinkage stresses in restrained element.

4.4 THERMAL SHRINKAGE

4.4.1 Mechanism

Massive concrete structures frequently occur in huge civil and structural engineering projects, such as dams, locks, tunnels, storage tanks, quay walls, foundation slabs, and concrete armour units. Due to the heat of hydration, thermal stresses occur during hardening, possibly leading to early-age thermal cracking (De Schutter 1996). The problem of early-age thermal cracking in massive concrete structures is not new. Early examples mentioning severe damage ascribed to thermal stresses occur in the literature such as the St. Francis dam in California in 1928. In more recent times, early-age thermal cracking has been reported in more slender structural elements made with high performance concrete containing very high cement contents.

Early-age thermal cracking can be subdivided in two categories: cracking due to internal restraint, and cracking due to external restraint. Although both categories typically occur simultaneously, they will be explained separately hereafter.

4.4.1.1 Early-age thermal cracking due to internal restraint

The main driving force for early-age thermal cracking in hardening concrete elements is the heat of hydration. Hydration of cement is an exothermic process. The produced hydration heat, depending on the type and amount of cement, causes a temperature rise within the concrete. As the heat conduction in concrete is relatively low (much lower than in steel), and as heat is exchanged with the environment at the exposed surfaces, the core of the concrete element will show a higher temperature than the surface zone. The environmental temperature as well as the casting temperature of the fresh concrete will have an important influence on the observed difference. Furthermore, the temperature gradient between core and surface will be further enlarged due to the fact that the hydration process is accelerated at higher temperatures.

In the case where the concrete core and surface are able to deform freely, different thermal deformations (expansions) would be obtained, as shown in Figure 4.8. However, as the concrete element is one solid block, free deformations of the different zones are not possible. The larger free thermal expansion of the element core will be restrained by the smaller free thermal expansion of the concrete surface zone. As a result, an equilibrium deformation is obtained, inducing thermal compressive stresses in the core and tensile stresses in the surface zone. The magnitude of the thermal stresses at this stage heavily depends on the Young's modulus and on the creep and relaxation behaviour of the hardening concrete. Thermal cracks can be formed in the surface zone when the tensile strength of the hardening concrete is not sufficient to withstand the thermal tensile stresses. As the Young's modulus of hardening concrete is developing relatively faster than the strength, a high cracking risk can occur.

At a later stage within the hardening process, the hydration process will slow down, and the concrete element will start to cool down. As the core of the element will have to cool down more than the surface zone, a stress inversion will be obtained, as shown in Figure 4.8. Tensile stresses will

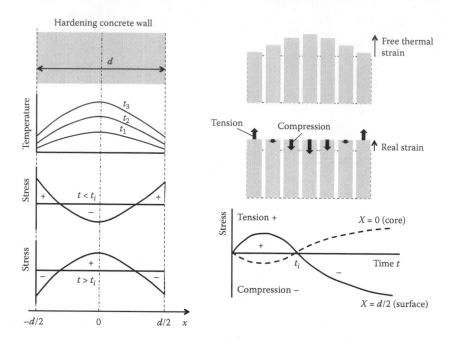

Figure 4.8 Early-age thermal stress formation in a massive hardening concrete element.

now be induced in the core of the element, while the surface zone is now in compression. An important element for a good understanding of the stress evolution in the hardening concrete element is the fact that the mechanical properties of the hardening concrete are evolving with time and also can be different from location to location. This will be further explained in detail.

Following the phenomenon as described in the previous paragraphs, early-age thermal cracking can occur during two stages: during the heating phase and during cooling down. During heating, when the concrete core is at higher temperatures, thermal cracking can occur in the surface zone. The cracking risk can be substantially increased by inappropriate demoulding, causing a sudden decrease in surface temperature, and a sudden increase in thermal gradient. In a cube-like concrete element, this can lead to a cross-like crack pattern as illustrated in Figure 4.9 (Mustard 1965). This crack pattern was observed in about 1% of grooved cube massive concrete armour units in the harbour of Zeebrugge, Belgium, in the 1980s, as shown in Figure 4.10.

It is also possible that early-age thermal cracking only occurs during the cooling phase, caused by excessive tensile stresses within the core of the element. However, this case is difficult to diagnose, as the internal cracks will not be visible. Numerical simulation could be helpful to verify crack formation at this stage.

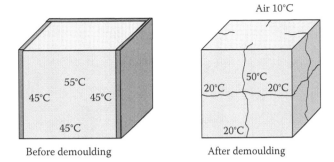

Figure 4.9 Cross-like crack pattern in hardening concrete cube due to thermal stresses in combination with inappropriate demoulding.

Figure 4.10 Early-age thermal cracks in grooved cube concrete armour unit.

4.4.1.2 Early-age thermal cracking due to external restraint

Cracks of a different type can occur when the thermal deformation of the hardening concrete element is restrained by a previously cast adjacent element or foundation. Also in this case, the heat of hydration is the driving force. Especially during the cooling phase, thermal stresses can lead to severe through-going cracks, which can seriously affect the water tightness of tunnel walls.

Consider a hardening concrete wall, cast on a previously cast floor or foundation, as illustrated in Figure 4.11. During the heating phase, the newly cast wall shows a thermal expansion, which is at least partly restrained by the already hardened floor or foundation. As a result, compressive stresses are

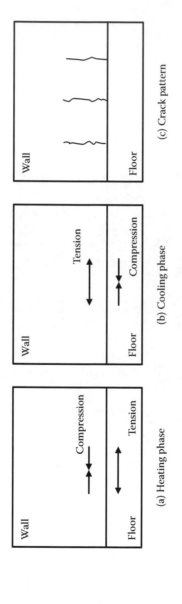

Figure 4.11 Early-age thermal cracking due to external restraint: example of a hardening wall on a previously cast floor or foundation.

induced in the hardening wall and tensile stresses in the floor. At this point, the cracking risk will be low, as the floor already has fully developed tensile strength and the hardening wall is in compression. (However, when the wall thickness is high, a thermal gradient can occur in the wall itself, possibly leading to surface cracks in the wall, as explained in the previous case of internal restraint.)

A more dangerous situation is obtained during the cooling phase. The thermal contraction of the hardening wall will be restrained by the floor. At this stage, tensile stresses occur in the wall and compressive stresses in the floor. Through cracks can occur in the hardening wall, going from one surface to the other. This type of crack formation can have important consequences for water tightness and durability of the wall.

In case no thermal cracking occurs, thermal stresses will remain in the hardened concrete element as eïgenstresses or self-stresses. These eïgen-stresses have to be duly considered as the initial state when further mechanically loading the element.

4.4.1.3 Importance of evolving mechanical properties

In order to better understand the phenomenon of early-age thermal cracking in hardening concrete elements, it is important to study the role of the evolving mechanical properties. By means of some elementary theoretical example, it will be illustrated here that a Young's modulus which varies in space and in time (due to the evolution of the hydration process) has a specific influence on the occurrence of and possible continual thermal stresses.

Consider an elastic body consisting of two halves as illustrated in Figure 4.12. Both halves have the same time-independent Young's modulus, E, but undergo a different temperature variation at time t, namely $\Delta\theta_1$ and $\Delta\theta_2$. It is supposed that the deformation of the two halves can only happen in one dimension (no transversal expansion, no rotation). It is further supposed that both halves have a unit length and a unit section, so that equilibrium equations can further be based on stresses and strains directly.

If both halves were able to deform independently, the situation shown in Figure 4.12b would occur, showing different thermal dilation and no stresses. The thermal strains ε_1 and ε_2 can be obtained by:

$$\varepsilon_1 = \Delta\theta_1 \cdot \alpha_t \tag{4.3}$$

$$\varepsilon_2 = \Delta\theta_2 \cdot \alpha_t \tag{4.4}$$

in which α_t is the coefficient of thermal expansion (CTE), which is considered to be the same for both halves.

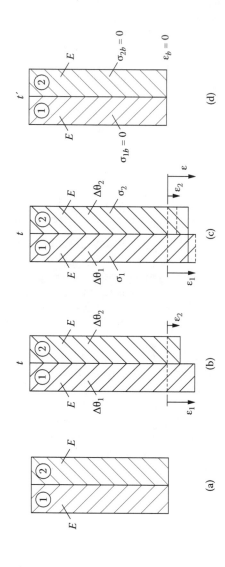

Figure 4.12 Evolution of thermal stresses: case of constant Young's modulus.

As the two halves are fully connected to each other, these free thermal strains cannot occur. Due to compatibility of strains, an overall strain ε will occur (Figure 4.12c), now leading to thermal stresses σ_1 and σ_2 in the two halves, which can be calculated as follows (compressive stresses taken as positive):

$$\sigma_1 = E(\varepsilon_1 - \varepsilon) \tag{4.5}$$

$$\sigma_2 = E(\varepsilon_2 - \varepsilon) \tag{4.6}$$

By expressing equilibrium ($\sigma_1 + \sigma_2 = 0$), it can be shown that:

$$\varepsilon = \frac{\varepsilon_1 + \varepsilon_2}{2} = \frac{\Delta\theta_1 + \Delta\theta_2}{2} \cdot \alpha_t \tag{4.7}$$

and

$$\sigma_1 = -\sigma_2 = \frac{\Delta\theta_1 - \Delta\theta_2}{2} \cdot E \cdot \alpha_t \tag{4.8}$$

Suppose that at time t' the imposed temperature variations are withdrawn (Figure 4.12d). The final stress and strain conditions can then be obtained by:

$$\varepsilon_h = \varepsilon + \varepsilon' = \frac{\Delta\theta_1 + \Delta\theta_2}{2} \cdot \alpha_t + \frac{-\Delta\theta_1 - \Delta\theta_2}{2} \cdot \alpha_t = 0 \tag{4.9}$$

$$\sigma_{1b} = \sigma_{2b} = \sigma_1 + \sigma_1' = \frac{\Delta\theta_1 - \Delta\theta_2}{2} E\alpha_t + \frac{-\Delta\theta_1 + \Delta\theta_2}{2} E\alpha_t = 0 \tag{4.10}$$

In other words, no remaining stresses or strains are obtained when the two halves have returned to their original temperatures.

Now consider the more complicated case of a time and location dependent Young's modulus, as illustrated in Figure 4.13. It can be shown that the remaining strain and stresses after returning to the original temperature are obtained by:

$$\varepsilon_b = \varepsilon + \varepsilon' = \left[\frac{E_1\Delta\theta_1 + E_2\Delta\theta_2}{E_1 + E_2} - \frac{E_1'\Delta\theta_1 + E_2'\Delta\theta_2}{E_1' + E_2'} \right] \cdot \alpha_t \tag{4.11}$$

$$\sigma_{1b} = -\sigma_{2b} = \left[\frac{E_1 E_2}{E_1 + E_2} - \frac{E_1' E_2'}{E_1' + E_2'} \right] \cdot \frac{\Delta\theta_1 - \Delta\theta_2}{2} \cdot \alpha_t \tag{4.12}$$

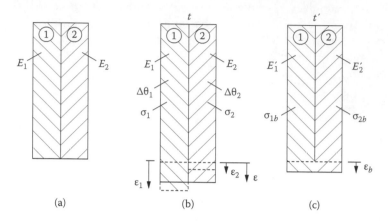

Figure 4.13 Evolution of thermal stresses: case of evolving Young's modulus.

In other words, a residual strain and residual stresses (so-called eïgen-stresses) are obtained in the two halves of the element.

This case is more representative for massive hardening concrete elements, where a time and location dependent Young's modulus indeed occurs. The example nicely illustrates the occurrence of eigenstresses which remain in the hardened concrete element. However, hardening concrete is much more complicated, because the material is not behaving in a purely elastic way. As a result, a very complex stress distribution is obtained in hardening concrete elements, which can only be accurately estimated by means of advanced numerical techniques (De Schutter 1996, De Schutter 2002).

4.4.2 Influencing parameters

4.4.2.1 Parameters related to the concrete composition

As the heat of hydration is the main driving force for early-age thermal cracking, the cement type is of major importance. Ordinary Portland cement typically produces more heat of hydration than blended cements such as blast furnace slag cement. Finer cements generally show higher heat production rates than coarser cements. Puzzolan or mineral additions can help reduce the heat of hydration. The effect of the additions on the hydration reaction of the cement should be carefully considered, especially in the case of high powder contents such as in self-compacting concrete (De Schutter et al. 2008).

A higher cement content will lead to an increased heat production. This is the reason why high strength concrete elements might show some risk of early-age thermal cracking, even when the element might not seem very massive.

The water/cement ratio does not seem to influence significantly the heat production rate during hardening. However, as a higher water/cement ratio will lead to a higher ultimate degree of hydration, the total cumulated heat will be higher as well. Nevertheless, thermal stresses are governed more by the heat production rate during hardening and less by the total cumulated heat after a long time. It should be repeated, however, that a higher water/cement ratio will lead to lower strength values, and thus possibly to a higher cracking risk.

The nature of the aggregates can influence the evolving temperature fields during hardening by influencing the thermal properties of the concrete. The aggregate size, which can be quite large in very massive structures, can influence the paste volume of the concrete and thus the required cement content (see also Section 1.4.3 in Chapter 1).

Plasticizers can reduce the water/cement ratio and thus increase the concrete strength, possibly reducing the early-age thermal cracking risk. However, a lower water/cement ratio will lead to a higher autogenous shrinkage and a higher cement content might also be needed to maintain sufficient workability. This can make the situation very complex in the case of high strength concrete with a very low water/cement ratio.

4.4.2.2 Geometrical parameters

The massivity of the concrete element is a very important parameter for the study of early-age thermal cracking. The massivity of an element is typically defined as the ratio of volume to surface. However, comparing different shapes or geometries, the massivity does not seem to enable quantitative comparisons. A more accurate parameter combining size and shape seems to be the equivalent thickness (De Schutter and Taerwe 1996b). The higher the equivalent thickness, the higher the risk of early-age thermal cracking.

4.4.2.3 Atmospheric parameters

At higher environmental temperature, the hydration process of the concrete will typically proceed faster. This will lead to an accelerated production of the heat of hydration and to higher temperature gradients in the hardening element. However, the influence of the environmental temperature also depends on the size of the element, the presence of the formwork, the casting temperature of the fresh concrete ... and thus should be verified in view of all these parameters.

Wind conditions can be very influential on the problem of early-age thermal cracking, especially immediately after demoulding. Also solar radiation can have a significant influence. Typically, a higher wind speed can lead to larger thermal gradients within the element. A more pronounced solar effect in combination with day and night cycles can also increase the early-age thermal crack risk.

4.4.2.4 Technological parameters during execution

During execution, the early-age thermal crack risk can be significantly influenced by the formwork conditions and by cooling measures. An insulated formwork will increase the maximum temperature in the hardening element, increasing the risk of early-age thermal cracking due to external restraint. However, it will decrease the thermal gradient inside the hardening element, thus reducing the risk of early-age thermal cracking due to internal restraint. By all means, the timing of demoulding should be carefully checked in case of massive hardening concrete elements. Inappropriate demoulding when still having elevated temperatures in the core of the element can lead to a substantial increase in thermal gradient and, thus, to an increased crack risk.

4.4.3 Mitigation

Considering the influencing parameters as mentioned in the previous section, some measures can be listed in order to avoid the risk of early-age thermal cracking in hardening concrete elements:

- A cement type with low heat of hydration can be considered. This can be a Portland cement with reduced C_3S content, designated as a low heat (LH) cement. As an alternative (and a more common solution in Europe), blended cements can be used, such as blast furnace slag cements.
- A partial cement replacement by puzzolan materials can also help reduce the heat of hydration. Puzzolan materials react more slowly than Portland cement and thus reduce the heat production rate. Puzzolans can help to reduce the cement content, while still maintaining long-term strength of the concrete. However, their influence on the long-term durability behaviour of the concrete should also be investigated, duly considering the type and nature of the puzzolan.
- A coarser aggregate type could be applied, considering the massivity of the element. This will lead to a lower required paste volume, which is helpful in further reducing the cement content and, thus, the heat of hydration.

- Retarders are only helpful in controlling the early-age crack risk when they lead to a slowing down of the entire hydration process. Many commercially available retarders only influence the dormant period, delaying the setting time. However, after the dormant period, the hydration reaction follows the same rates as without a delayed setting. In this case, the retarder will not be helpful to decrease the risk of early-age thermal cracking.
- Reinforcement will not significantly influence the thermal field in hardening of massive concrete elements. The resulting thermal stresses will also not be influenced significantly. On the other hand, reinforcement can be helpful in reducing crack width and crack spacing in case early-age thermal cracking is occurring.
- Cooling measures can reduce the risk of early-age thermal cracking. Two types of cooling measures can be considered: cooling of the constituent materials before mixing and casting, or incorporating cooling pipes in the concrete element in order to evacuate (part of) the heat of hydration during hardening.
- An appropriate choice of formwork type (insulating or not), in combination with an appropriate demoulding strategy can also be helpful in avoiding early-age thermal cracking. This should, however, be studied case by case.

For a more detailed study of practical measures to avoid early-age thermal cracking in massive hardening concrete elements, reference is made to the literature (De Schutter 1996).

4.4.4 Example

For the construction of a harbour dock in Antwerp, Belgium, a new type of non-reinforced massive concrete quay wall has been designed. While for previous docks reinforced L-shaped walls have typically been constructed, the new type of quay wall (Figure 4.14) has a so-called J-shape, a total height of about 30 meters, a maximum thickness of about 20 meters, and contains no reinforcement at all. The self-weight of the quay wall is sufficient to withstand the ground pressure and the shape makes sure that compressive as well as tensile stresses within the concrete remain below acceptable levels. The volume of the quay wall is about 300 m³/m. The unique combination of geological conditions and the possibility to build the quay wall in open-cut, enabled the construction of the J-shaped type of quay wall on this specific location in Antwerp (De Schutter and Vuylsteke 2004).

Due to the massivity of this quay wall, a major issue however is the problem of early-age thermal cracking. An important aspect herewith is the casting procedure. It was studied whether casting in several layers would reduce the risk of early-age thermal cracking. Finite element simulations

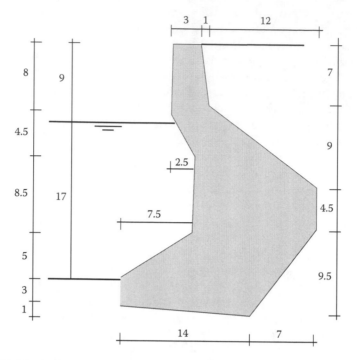

Figure 4.14 Massive J-type quay wall.

were performed, implementing all relevant material properties (thermal and mechanical), and considering relevant boundary conditions (also thermal and mechanical).

Casting the massive quay wall in different horizontal layers seemed inadvisable because some kind of thermal shock was occurring between the newly cast layer (temperature of the fresh concrete) and the previous layer (higher temperature due to heat of hydration). This thermal shock was causing possibly severe damage in the contact zone between the successive layers, as illustrated in Figure 4.15.

Cooling of the fresh concrete, e.g. by cooling of the aggregates or the mixing water, was shown to be a very effective measure in order to avoid thermal cracking at the surface of the quay wall. However, when casting in layers, the damage in the contact zone between layers could not be avoided by cooling the fresh concrete.

An optimal approach seemed to be the casting of the quay wall in one operation, from bottom to top, combined with cooling of the fresh concrete. In this way, the internal damage can also be avoided. However, some logistic problems have to be solved concerning cooling and mixing of the concrete in considerable quantities during a short time period.

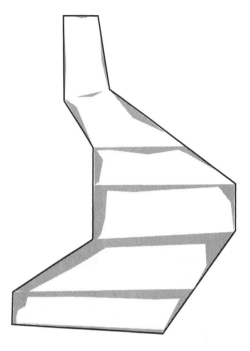

Figure 4.15 Damage zones due to thermal stresses in massive quay wall when casting in 6 different layers.

REFERENCES

Almudaiheem J.A. and Hansen W. (1987) 'Effect of specimen size and shape on drying shrinkage of concrete', *ACI Materials Journal,* 84, 985–994.

Baroghel-Bouny V., Mounanga P., Khelidj A., Loukili A., Rafaï N. (2006) 'Autogenous deformation of cement pastes. Part II. W/C effects, micro-macro correlations, and threshold values', *Cement and Concrete Research* 36, 123–136.

Beltzung F., Wittmann F.H. and Holzer L. (2001) 'Influence of composition of pore solution on drying shrinkage', Proceedings of the 6th International Conference on creep, shrinkage and durability mechanics of concrete and other quasi-brittle materials (*ConCreep* 6), Boston, USA, 39–48.

Bentur A. (Ed.) (2003) 'Early Age Cracking in Cementitious Systems', RILEM, Report 25, ISBN: 2-912143-33-0, pp. 388.

Bisschop J. (2002) 'Drying shrinkage microcracking in cement-based materials', Doctoral Thesis, Delft University of Technology, the Netherlands, pp. 198.

Bissonnette B., Pierre P. and Pigeon M. (1999) 'Influence of key parameters on drying shrinkage of cementitious materials', *Cement and Concrete Research,* 29, 1655–1662.

Bjøntegaard Ø. (1999) 'Thermal dilation and autogenous deformation as driving forces to self-induced stresses in high-performance concrete', PhD thesis, Norwegian University of Science and Technology, Trondheim, Norway.

Craeye B., De Schutter G., Desmet B., Vantomme J., Heirman G., Vandewalle L., Cizer Ö., Aggoun S. and Kadri E.H. (2010) 'Effect of mineral filler type on autogenous shrinkage of self-compacting concrete', *Cement and Concrete Research*, 40, 908–913.

Craeye B., Geirnaert M. and De Schutter G. (2011) 'Super absorbing polymers as an internal curing agent for mitigation of early-age cracking of high-performance concrete bridge decks', *Construction and Building Materials*, 25, 1–13.

De Schutter G. (1996) 'Fundamental and Practical Study of Thermal Stresses in Hardening Massive Concrete Elements' (in Dutch), Doctoral Thesis, Magnel Laboratory for Concrete Research, Ghent University, Belgium, pp. 364.

De Schutter G. and Taerwe L. (1996) 'Degree of hydration based description of mechanical properties of early age concrete', *Materials and Structures*, 29, 335–344.

De Schutter G. and Taerwe L. (1996b) 'Estimation of the early age thermal cracking tendency of massive concrete elements by means of the equivalent thickness.', *ACI Materials Journal*, 403–408.

De Schutter G. (2002) 'Finite element simulation of thermal cracking in massive hardening concrete elements using degree of hydration based material laws', *Computers and Structures*, Vol. 80, Issue 27-30, p. 2035–2042.

De Schutter G. and Vuylsteke M. (2004) 'Minimisation of early age thermal cracking in a J-shaped non-reinforced massive concrete quay wall', *Engineering Structures*, 26, 801–808.

De Schutter G., Bartos P., Domone P. and Gibbs J. (2008) *'Self-Compacting Concrete'*, Whittles Publishing, Caithness, UK, CRC Press, Taylor & Francis Group, Boca Raton, USA, ISBN 978-1904445-30-2, USA ISBN 978-1-4200-6833-7, pp. 296.

Esping O. (2008) 'Effect of limestone filler BET(H20)-area on the fresh and hardened properties of self-compacting concrete', *Cement and Concrete Research*, 38, 938–944.

Hansen T.C. (1986) 'Physical structures of hardened cement paste. A classical approach', *Materials and Structures*, 19, 423–431.

Hobbs D.W. and Mears A.R. (1971) 'The influence of specimen geometry upon weight change and shrinkage of air-dried mortar specimens', *Magazine of Concrete Research*, 23, 89–98.

Kovler K. and Jensen O.M. (Eds.) (2007) 'Internal Curing of Concrete', State-of-the-art report of RILEM Technical Committee 196-ICC, RILEM Report 41, RILEM Publications S.A.R.L., ISBN 978-2-35158-009-7, pp. 140.

Koenders E.A.B. (1997) 'Simulation of volume changes in hardening cement-based materials', PhD thesis, Technical University of Delft, the Netherlands.

Loukili A., Chopin D., Khelidj A., Le Touzo J.-Y. (2000) 'A new approach to determine autogenous shrinkage of mortar at an early age considering temperature history', *Cement and Concrete Research* 30, 915–922.

Miao C., Tian Q., Sun W. and Liu J.P. (2007) 'Water consumption of the early-age paste and the determination of "time-zero" of self-desiccation shrinkage', *Cement and Concrete Research*, 37, 1496–1501.

Mindess S. and Young J.F. (1981) 'Concrete', Prentice-Hall.

Mustard H. (1965) 'Mass concreting principles applied to massive structural members', *ACI Journal*, June, 651–660.

Neville A.M. and Brooks J.J. (2010) '*Concrete Technology*', Second Edition, Prentice Hall, Pearson, ISBN 978-0-273-73219-8.

Robeyst N., Gruyaert E., Grosse C.U., De Belie N. (2008) 'Monitoring the setting of concrete containing blast furnace slag by measuring the ultrasonic p-wave velocity', *Cement and Concrete Research* 38, 1169–1176.

Tazawa E. (Ed.) (1999) 'Autogenous shrinkage of concrete', Proceedings of the International Workshop organized by JCI, Hirshima, E&FN Spon, London, ISBN 0-419-23890-5, pp. 406.

Therrien J., Bissonnette B. and Cloutier A. (2000) 'Early-age evolution of the mass transfer properties in mortar and its influence upon ultimate shrinkage', Proceedings of the International RILEM Workshop on Shrinkage of Concrete', Paris, France.

Wittmann F.H. (2001) 'Mechanism and mechanics of shrinkage', Proceedings of the 6th International Conference on creep, shrinkage and durability mechanics of concrete and other quasi-brittle materials (ConCreep 6), Boston, USA, 3–12.

Chapter 5

Actions during service

5.1 MECHANICAL ACTIONS

5.1.1 Direct loading and impact

Concrete structures or structural elements are designed to take up external mechanical loading. In modern design codes based on ultimate limit state design, it is considered that concrete in tension will crack, leaving only the reinforcement to carrying the tensile loading. This does not mean that concrete structural elements under service load will always show cracks. However, it does mean that if cracks occur in concrete structures, the structural load-bearing capacity of the structure is not immediately at risk. Of course, excessive cracking in concrete structures is neither generally expected nor accepted. And even small cracks could impair long-term concrete durability because of the accelerated ingress of aggressive substances.

Cracks due to direct mechanical loading can be easily recognized by looking at the crack pattern. Different load conditions will lead to typical crack patterns, as illustrated in Figure 5.1. In laboratory conditions, when the structural performance of concrete elements is experimentally tested, a fully developed crack pattern can be obtained when loading up to failure. However, in real practice, when the concrete element is well designed, the number of cracks and the crack widths will be rather limited. Excessive cracking might indicate inappropriate design or unexpected overloading of the structure.

Crack widths will depend on several parameters such as the bond between concrete and steel reinforcement, cover thickness, distance between reinforcing bars, and diameter of the reinforcing bars. Although considerable agreement exists concerning the principles of crack simulations and estimations, different codes will typically predict different crack widths. Differences also exist in prescribed maximum allowable crack widths according to different design codes. For reinforced concrete, crack widths up to 0.3 mm are widely accepted in cases of non-severe exposure conditions. In the case of a more aggressive environment, allowable crack widths are typically reduced to 0.2 or even 0.1 mm.

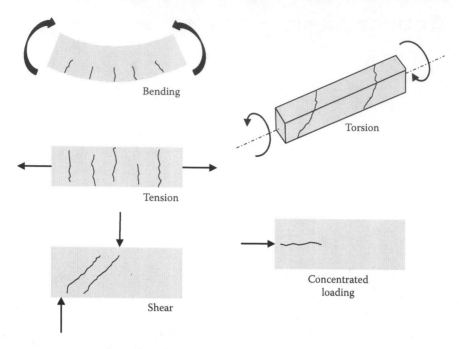

Figure 5.1 Crack patterns due to direct loading.

Significant damage to concrete structures can also be caused by direct loading due to earthquakes. In cases where no special provisions have been taken against loading by earthquakes, structures can totally collapse, causing many casualties. Many countries situated in active earthquake zones have special requirements concerning the design of concrete structures. Even in these countries, earthquakes can still severely damage concrete structures, as illustrated in Figure 5.2.

Another case of direct loading is impact. This can be due to explosions (e.g. industrial accidents or terrorist attacks) or traffic accidents. A typical example of this is shown in Figure 5.3, which shows damage to a concrete girder after it was hit by a truck that failed to respect the height limits. In these cases, the remaining structural capacity of the girders must be carefully examined, considering possible damage to the prestressing wires or cables in the case of prestressed bridge girders.

In reinforced concrete slabs of road bridges with limited thickness compared to the span length, fatigue failure due to repeated loading can lead to a two-directional crack pattern as shown in Figure 5.4. In the early stages of this deterioration process, cracks occur perpendicular to the bridge axis. This is linked to a lower reinforcement ratio in the transversal direction of the bridge. With increased load repetition, the crack pattern becomes two-directional. Cracks will grow, possibly causing a push-out of the concrete.

Figure 5.2 Damage to concrete columns due to earthquake loading (Internet photo).

Figure 5.3 Impact damage to concrete girder by truck.

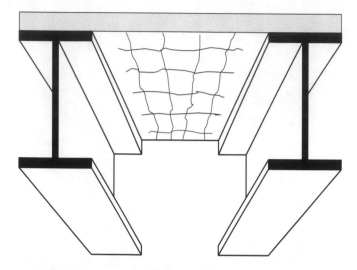

Figure 5.4 Fatigue damage in a thin floor slab of a reinforced concrete bridge.

5.1.2 Differential settlements

In structures which are statically indeterminate (such as a continuous bridge girder on three or more supports), differential settlements can lead to cracking as illustrated in Figure 5.5. The resulting crack patterns can be compared with the crack patterns caused by direct loading. Indeed, a differential settlement of a statically indeterminate structure will have the same effect as an external load.

Consider as an example the first situation shown in Figure 5.5. Settlement of the middle support can be compared with the situation of the entire beam only supported by the two end supports and loaded by a vertical downward force at the location of the middle support. The resulting crack pattern is thus completely similar to the crack pattern of a beam in bending.

Figure 5.6 shows the possible effect of differential settlements in the case of a concrete brick wall. The lines of damage are very similar to the crack pattern in a continuous reinforced concrete wall loaded in shear.

5.2 PHYSICAL ACTIONS

5.2.1 Frost damage

5.2.1.1 Mechanism

Concrete is a porous material containing pores with a multitude of sizes ranging from nanometre to millimetre (Neville and Brooks 2010).

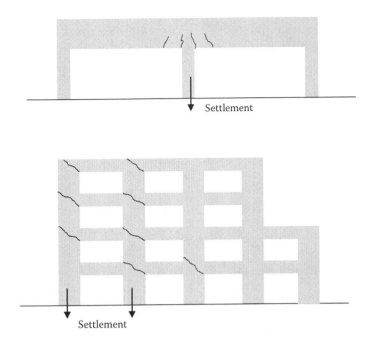

Figure 5.5 Damage due to differential settlements.

Figure 5.6 Effect of differential settlements on a concrete brick wall.

These pores typically are partially filled with water containing a variety of ions, the so-called pore solution. When concrete is exposed to freezing temperatures, the pore water can start freezing which will result in ice formation and possibly in damage to the concrete, which is called frost damage or internal frost damage. When de-icing salts are applied to the concrete surface, as is common practice on concrete roads in many countries with cold winter climates, the damage mechanism is somewhat different, resulting in so-called salt scaling. This will be further detailed in Section 5.2.2. First, internal frost damage (without de-icing salts) will be explained.

To explain internal frost damage in concrete, the expansion associated with the transformation of water into ice while freezing is an essential element. Nevertheless, different theories exist, giving different views on the real mechanisms occurring during freeze–thaw cycles in concrete.

5.2.1.1.1 Hydraulic pressure

Based on the expansive ice formation when water is freezing, representing a 9% volume increase, Powers (1945) proposed the hydraulic pressure theory to explain internal frost damage. Ice formation is initiated first in the larger pores within the capillary pore structure, expelling pore water from the freezing pores due to the aforementioned volume increase. Depending on the growth rate of the ice, which is a function of the cooling rate and the level of undercooling before nucleation occurs (Sun and Sherer 2010), and depending on the pore structure of the cement matrix, significant hydraulic pressures could occur, leading to tensile stresses in the concrete, possibly reaching the tensile strength and causing frost damage.

Figure 5.7 illustrates the principles of the hydraulic pressure theory. Expansive ice formation is occurring in a capillary pore, and is pressurizing the non-frozen water. Due to this pressure, the water will be expelled

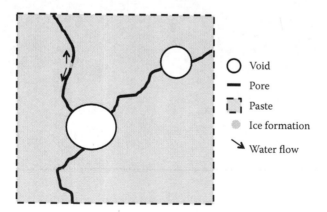

Figure 5.7 Water flow due to ice formation in capillary pores.

from the capillary, finding its way to air voids. When enough air voids are available within a short distance, the hydraulic pressure can be easily relaxed (see Section 5.2.1.3: application of air entrainment to improve frost resistance). Otherwise, the hydraulic pressure can result in damage and the cracking of the freezing concrete.

Although the hydraulic pressure theory seems to logically explain the occurrence of internal frost damage, Powers observed phenomena which could not be easily explained (Powers 1953, Valenza and Scherer 2007b). While reducing the air void spacing, i.e. facilitating the flow of pore water when expelled by the expansive ice formation, a contraction of the concrete is obtained larger than the thermal contraction, instead of a mere reduction of the expansion caused by the hydraulic pressure. Furthermore, when cooling is stopped, no relaxation of the hydraulic pressures seems to occur in air-voided concrete, while the hydraulic pressure theory would predict a stress relaxation due to migration of water to the available air voids. After further analysis, Powers finally withdrew his own hydraulic pressure theory (Powers 1975). Nevertheless, the capillary pressure theory is useful to explain frost damage caused by ice formation in capillary pores. In some cases, when water-filled pores are surrounded by growing ice crystals, the resulting fluid pressure indeed is believed to cause frost damage (Chatterji 2003).

5.2.1.1.2 Crystallisation pressure

It is important to know that in fine pores the freezing point is reduced because the ice crystals have a high surface-to-volume ratio. At the normal freezing temperature, a minimization of Gibbs energy will not be obtained while forming small ice crystals because the energy gain due to solidification is counteracted by the energy of the interface between the small ice crystals and the surrounding water (Valenza and Scherer 2007b). Based on energy minimization, a relationship can be obtained between the freezing point and the largest spherical crystal that can be formed in a pore with a certain radius, taking into account a small layer (about 0.9 nm thick) of unfrozen water which remains between ice crystals and pore wall (Valenza and Scherer 2007b). Without going into the details of this relationship, it can be concluded that the freezing point is reduced by 2°C for a capillary with a radius of about 33 nm, by 5°C for a radius of about 13 nm, and by 10°C for a radius of about 7 nm (Valenza and Scherer 2007b).

As a result of the relationship between a freezing temperature and capillary radius, within a saturated cementitious material ice formation will be first initiated in the larger pores or voids (see Figure 5.8). In the meantime, the water in the smaller pores will remain liquid, in a state of super cooling. This results in a thermodynamic disequilibrium, which acts as a driving force for the water to move from the smaller capillary pores to the larger voids where ice crystals are being formed (Mehta and Monteiro 2006).

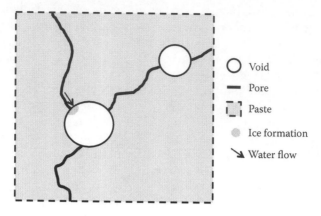

Figure 5.8 Water flow due to ice formation in air void.

When insufficient space is left for further ice formation, internal pressures will be initiated, causing expansion and possibly cracking.

Following from the explained mechanism, expansion will be caused in the area of the freezing void, while contraction is caused in other regions by the local loss of water due to migration towards the ice crystals. In concrete without air entrainment, the net result will be an expansion. In air-entrained concrete, overall contraction will be noticed (Mehta and Monteiro 2006). Thus, the crystallisation theory is able to explain the previously mentioned observation, which could not be explained by the hydraulic pressure theory. The same holds for the observation that the pressure is not relaxed when cooling is stopped.

When the temperature of the freezing concrete is further reduced, ice crystals will start growing into the smaller capillary voids and pores. Damage can further occur due to the expansion of the ice. Although expansion during ice formation is, according to the crystallisation theory, responsible for damage formation, the mechanism is not equivalent to the usual mechanism of hydraulic pressure (Valenza and Scherer 2007b). To make this more clear, reference is made to Beaudoin and McInnis (1974) who showed that the expansion of freezing cement paste can be observed even in the case of benzene (a liquid showing contraction when freezing) as pore fluid instead of water. Thus, the driving force for internal frost damage seems to be the thermodynamic disequilibrium, which moves the water toward the growing ice crystals, making them grow further. Due to the resulting crystallisation pressure, the concrete can be damaged.

In practice, concrete structures most often will not be fully saturated. While freezing, ice, water and water vapour can simultaneously exist within the concrete. Every frost–thaw cycle will cause moisture movement in the material, making it a very dynamic process (Setzer 2003, Setzer et al. 2004).

5.2.1.1.3 Osmotic pressure

While the temperature is decreasing, ice formation first occurs in the larger pores, as explained before. Due to the ice formation, mainly involving pure water, the alkalinity of the remaining non-frozen water within the larger pores will increase. In order to reach thermodynamic equilibrium, pore water with lower alkalinity will move from the smaller pores to the larger pores where ice formation occurs. As a result, the alkalinity of the non-frozen water near the ice crystals will decrease, facilitating further ice formation. Otherwise, due to the diffusion of solution towards the zones with higher alkalinity, the fluid pressure will rise here, until equilibrium is obtained with the difference in osmotic pressure between the zones with higher and lower alkalinity. However, based on some calculations, Valenza and Scherer (2007b) showed that the osmotic pressure is very unlikely to cause concrete damage. Nevertheless, the osmotic pressure could be considered as an additional phenomenon, adding to the more important crystallisation pressure.

5.2.1.2 Influencing parameters

5.2.1.2.1 Degree of saturation

The degree of saturation of the concrete is calculated as the amount of free water present in the pores, relative to the maximum amount of water when all the pores would be filled. The degree of saturation of the concrete is an important parameter regarding frost damage. Dry concrete, of course, will not show any damage when exposed to freezing because no ice will be formed. In partially saturated concrete, a certain volume of the pores will contain no water. This empty pore volume will provide the possibility that ice will expand or pressurized water will escape, releasing pressure.

Considering the fact that a 9% volume increase is noticed during ice formation, it would be reasonable to expect that frost damage will not occur in concrete with a degree of saturation below about 90%. Experimental results confirm the existence of a critical saturation degree below which no significant frost damage will occur, although proposed values range between 85% and 90%.

5.2.1.2.2 Air void system

Besides the amount of free expansion volume, which can be deduced from the degree of saturation, the spatial distribution of this expansion volume is also important. The available expansion volume is only effective when it is easily and rapidly accessible for pressurized water or expanding ice. In order to improve frost resistance, air-entraining agents are quite often used in concrete, providing a large amount of small air voids evenly distributed within the cementitious system. A key parameter concerning the successful application of air-entraining agents is the spacing factor, which is related to the maximum distance to the nearest air void. Values below 200 µm are typically considered to provide adequate frost resistance.

5.2.1.2.3 Water/cement ratio

A low water/cement ratio will lead to the reduced permeability of the concrete. It could be reasoned that this will lead to higher hydraulic pressures when ice formation is initiated because the pressurized water cannot easily escape. In practice, however, a lower water/cement ratio will lead to a higher frost resistance. Due to the reduced permeability, the critical saturation degree will almost never be reached in real situations. Furthermore, the reduced pore volume results in a reduced amount of freezable water in the concrete. As a result, the tensile stresses in the concrete matrix, as caused by a limited amount of freezing water, will not cause significant damage in the case of a low water/cement ratio.

5.2.1.2.4 Strength

As a lower water/cement ratio leads to a higher concrete strength, it could be concluded that stronger concrete shows better frost resistance. However, this is not generally true. Durability requirements should not be narrowed to strength verification alone. Strength is an important parameter, but it is not sufficient to guarantee frost resistance. Air-entrained concrete shows lower strength than the corresponding concrete without air entrainment. Nevertheless, the air-entrained concrete shows better frost resistance (Mehta and Monteiro 2006).

5.2.1.2.5 Cement type

The type of cement influences the microstructure development, and thus the pore system and transport properties of the cementitious material. In this way, the cement type has a potential effect on frost resistance. However, a more significant influence is noticed through the interaction between carbonation and frost resistance. As an example, concrete based on blast furnace slag cement will be more vulnerable to carbonation. Furthermore, carbonation of slag concrete will lead to an increased porosity and permeability, while the opposite is noticed in the case of Portland cement concrete. As a result, carbonated slag concrete will show more pronounced frost damage. Inadequate curing of concrete structures at early age can significantly increase the carbonation rate in slag concrete and, thus, also further impair frost resistance. This should be well considered when evaluating the frost resistance performance of slag concrete in lab conditions, having perfect curing and no carbonation.

5.2.1.2.6 External parameters

Some external parameters will also influence frost damage to concrete structures.

- Freezing temperature: More significant frost damage can be noticed for lower temperatures.
- Freezing rate: In the case of a higher freezing rate (faster temperature drop), more damage can be expected.
- Frost–thaw cycles: Significant frost damage does not always occur during the first frost–thaw cycle, but rather after a number of repeated cycles. This is related to further water movements (Setzer 2003, Setzer et al. 2004) and possibly to some mechanical fatigue behaviour in the concrete matrix.

5.2.1.3 Mitigation

Frost damage in concrete can easily be reduced or prevented by reducing the water/cement ratio and by adding an air-entraining agent. Both measures are typically considered in concrete standards (see also Chapter 1). Prescribed maximum water/cement ratios typically range between 0.45 and 0.50, depending on the frost risk (linked to the degree of saturation). When applying air-entraining agent, additional air content ranging from 4% to 6% is typically the goal. The air voids need to have a diameter between 100 µm and 500 µm, and the spacing factor should be smaller than 200 µm. It is to be considered that for each percent of entrained air, the compressive strength of the concrete will be reduced by about 5%.

5.2.1.4 Example

As a somewhat special example of frost damage, reference can be made to hollow core slabs exposed to weather during construction. Water can accumulate within the cores. Upon freezing, the hollow core slabs can be severely damaged (Buettner and Becker 1998).

5.2.2 Frost in combination with de-icing salts (salt scaling)

5.2.2.1 Mechanism

De-icing salts are often used on concrete roads and pathways in order to keep traffic going during hard winter times. Typically, the de-icing salts are based on $NaCl$, $CaCl_2$, and $MgCl_2$ or on different kinds of acetate. The application of de-icing salts can lead to a particular damage pattern visible on the concrete surface, described as 'scaling'. Flakes of material come loose on the top surface, finally showing a patchy damage pattern.

In order to explain the damage caused by frost in combination with de-icing salts, reference is often made to the same theories as valid for classical frost damage: hydraulic pressure, crystallisation pressure, and osmotic pressure. Nevertheless, the role of the de-icing salts could be better understood by considering the salt concentration gradient theory, or by the 'glue-spall' theory. This latter now seems to be the most widely accepted theory to explain the specific damage mechanism related to frost in combination with de-icing salts, also referred to as salt scaling. Other theories exist, e.g. the thermal shock theory, and the precipitation and growth of salt. However, these phenomena are not believed to contribute significantly to salt frost scaling (Valenza and Scherer 2007b).

5.2.2.1.1 Salt concentration gradient

When applying de-icing salts on a surface, a salt concentration gradient will be established toward the inner part of the concrete. This gradient will lead to a variation of the freezing point in the concrete, showing a lower freezing temperature for higher salt concentrations. Practically, the freezing temperature will be lower at the concrete surface and will increase when going deeper into the concrete. This is illustrated in Figure 5.9.

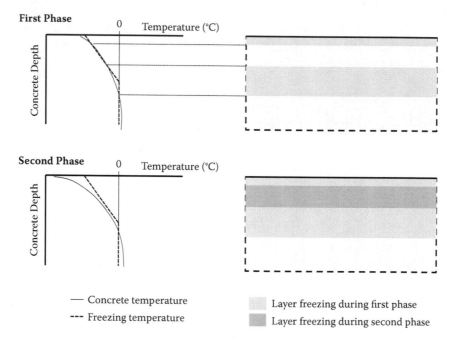

Figure 5.9 Effect of salt concentration gradient on freezing temperature and salt scaling.

As a consequence of the spatial evolution of the freezing temperature, the top layer and a more inward layer of the concrete can be frozen, while in an intermediate layer the water (salt solution) will still be liquid. When the temperature further drops, this intermediate layer can also start freezing. However, following the hydraulic pressure theory, at this stage the hydraulic pressure cannot be relaxed because the water cannot enter the already frozen top and bottom layers. As a result, the already frozen top layer will be (locally) pushed off, causing a scaling damage pattern.

5.2.2.1.2 Glue spall theory

By compiling a critical review of the research concerning salt frost scaling, Valenza and Scherer (2007a) concluded that the salt concentration of the salty water layer on the concrete surface is more important than the salt concentration in the pore solution. Furthermore, a pessimum exists at a solute concentration of about 3% and no scaling occurs when the pool of solution is missing from the concrete surface. Valenza and Scherer (2007a) also conclude that the susceptibility to salt scaling is not correlated with the susceptibility to internal frost action, and that the ability to resist salt scaling is rather determined by the strength of the concrete surface. These findings cannot be explained by previously mentioned theories. As a result, the glue spall mechanism was proposed as the primary cause of salt scaling (Valenza 2005, Valenza and Scherer 2007b, Sun and Scherer 2010).

Due to the application of de-icing salts, a small layer of salt solution is formed on the horizontal concrete surface. When this solution layer freezes, it turns into a solid material. Upon further cooling, the thermal contraction of the ice layer is about five times higher than the thermal contraction of the concrete due to the large difference in the thermal expansion coefficient between ice (about $50.10^{-6}/°C$) and concrete (about $10.10^{-6}/°C$). The larger thermal contraction of the ice layer is restrained by the concrete substrate, causing tensile stresses within the ice layer. Depending on the salt concentration, cracks will occur in the ice layer. Following the principles of fracture mechanics, it can be shown that the cracks in the ice layer can penetrate into the concrete substrate, where they will bifurcate into a path parallel to the concrete surface (see Figure 5.10). As a result, small flakes will be removed from the concrete surface, leading to the well-known salt scaling damage pattern.

Following the glue spalling theory, salt scaling is an almost purely mechanical phenomenon involving issues of restraint (the thermal deformations of the ice layer are restrained by the concrete substrate) and fracture mechanics (cracks will occur in the ice layer, will propagate into the concrete substrate, and will bifurcate into a path parallel to the surface). Some physical issues also play a role, because the freezing temperature and the mechanical properties of the frozen solution depend on the salt concentration. Pure water ice is not expected to crack, brine ice formed from solutions

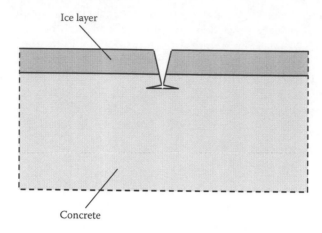

Figure 5.10 Mechanism of frost scaling according to glue spall theory.

of moderate concentration will crack, and highly concentrated solutions do not gain strength in the temperature range of interest (Valenza and Scherer 2007b).

Following the glue spalling theory, the beneficial role of air entrainment is explained in two ways (Valenza and Scherer 2007b). First, air entrainment reduces the amount of bleeding, producing a stronger concrete top surface. Furthermore, ice formation in the air voids forces the pore fluid to move toward the air voids (due to thermodynamic disequilibrium, see Section 5.2.1.1). This water movement from pores to larger voids leads to a slight contraction of the porous matrix, which helps reduce the damage caused by the thermal incompatibility between the ice layer on the concrete surface and the concrete substrate.

5.2.2.2 Influencing parameters

All parameters influencing frost damage of concrete (see Section 5.2.1.2) are typically considered to have a similar influence on frost salt scaling. However, following the glue spalling theory as concluded by Valenza and Scherer (2007b), 'Susceptibility to salt scaling is not correlated with susceptibility to internal frost action'. According to the glue spalling theory, the following parameters are of major importance for salt scaling.

5.2.2.2.1 Salt concentration

As already mentioned in Section 5.2.2.1, a pessimum exists at a solute concentration of about 3%. The salt concentration is an important parameter for the mechanical properties of the ice layer. For a low concentration,

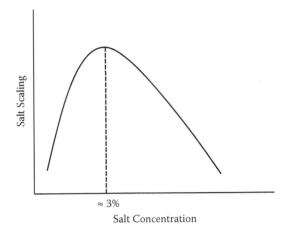

Figure 5.11 Influence of salt concentration on salt scaling.

the ice layer shows properties very similar to pure ice with a fairly high strength. In this case, the ice layer will not fracture when the temperature is further lowered and no significant scaling will occur. For a high concentration, the ice layer will not develop stiffness and strength within the relevant temperature range, so no significant stresses will be generated in the concrete. Maximum scaling damage will occur at a salt concentration of about 3% (see Figure 5.11).

5.2.2.2.2 Ice layer thickness

The higher the ice layer thickness, the higher the frost scaling damage. Indeed, when studying the thermal incompatibility between ice layer and concrete, it is clear that higher forces will be present in an ice layer with higher thickness. Upon the event of cracking, the crack will propagate somewhat further into the concrete, causing more spalling.

5.2.2.2.3 Quality of the concrete

The concrete properties near the top surface determine whether a crack forming in the ice layer will propagate into the concrete substrate. The higher the concrete strength, the lower the probability that cracks will propagate. As salt scaling is caused by a purely mechanical phenomenon, the concrete strength indeed seems to be decisive here. Nevertheless, it is also noticed that air entrainment is improving the resistance against salt scaling, even while the concrete strength is somewhat reduced. The reasons for the beneficial role of air entrainment in case of salt scaling have been explained in Section 5.2.2.1.2.

5.2.2.3 Mitigation

Although the damage mechanism in the case of salt scaling is totally different from the mechanism causing internal frost damage, better resistance can be obtained also in this case by reducing the water/cement ratio and by adding air entrainment.

5.2.2.4 Example

Frost scaling typically occurs on concrete roads and pavements exposed to de-icing salts. Figure 5.12 shows the resultant damage pattern in the case of a local concrete road exposed to de-icing salts in winter time. Due to traffic loading, the degradation of the concrete road can continue even after winter ends.

5.2.3 Shrinkage

Over the long term during service, drying shrinkage can lead to progressive crack formation in concrete structures. Although the process of drying shrinkage is much slower at the later age, the mechanism and influencing parameters are the same as for drying shrinkage during hardening. Reference is made to Section 4.3 'Drying shrinkage' in Chapter 4.

When the environment contains carbon dioxide, as is the case for the atmosphere, another type of shrinkage, the so-called *carbonation*

Figure 5.12 Salt scaling on a concrete road.

shrinkage, accompanies drying shrinkage. Carbonation is the reaction of carbon dioxide with hydration products present in the concrete. Due to carbonation, calcium hydroxide is converted to calcium carbonate. Other hydration products are decomposed as well (Neville 2010).

Carbonation of concrete is well-known for the initiation of reinforcement corrosion (see Section 5.4). However, carbonation also leads to a contraction of the affected concrete. The microstructure of the cement paste is modified due to the chemical carbonation reaction. In Portland-cement based systems, a decrease in porosity is obtained as well as a decrease in total volume. The latter causes the carbonation shrinkage, which is non-reversible.

In normal conditions, carbonation is a rather slow process proceeding from the exposed surface of the concrete. As the carbonation shrinkage is limited to the carbonated surface layer, a differential shrinkage is induced between surface and bulk. The carbonation contraction of the surface layer is restrained by the bulk of the concrete, potentially leading to map cracking or crazing. This crazing is only superficial as opposed to drying shrinkage cracks, which typically penetrate further into the bulk of the material.

While carbonation shrinkage is not a predominant damage mechanism for traditional concrete, it is far more important for autoclaved aerated concrete (AAC). Especially in the case of higher concentrations of CO_2, carbonation shrinkage of AAC will be more important than drying shrinkage. In this case, protective measures are recommended (Aroni 1993).

Carbonation of traditional concrete mainly leads to a reduced alkalinity and an increased corrosion risk of the rebars (see Section 5.4). This is different in AAC, a tobermorite-based material which is originally neutral or weakly alkaline (Matsushita et al. 2004a, 2004b, 2009). Steel reinforcement in AAC is typically coated with corrosion inhibitors because the tobermorite-based matrix has no corrosion preventive capacities (Aroni 1993). As a consequence, carbonation induced corrosion is not an issue for AAC. It is rather the modification of the microstructure due to carbonation, inducing significant shrinkage values, which is a predominant concern for AAC.

Carbonation of AAC leads to a decrease in strength, to an increase in deflection, and the growth of lattice-like cracking (Matsushita et al. 2009). For carbonation degrees below 25%, no significant carbonation shrinkage is noticed as the double-chain silicate anion structure of the tobermorite is well maintained (Matsushita et al. 2004b). However, carbonation shrinkage gradually increases for higher degrees of carbonation ranging from approximately 20% to 60%, most probably due to the decomposition of the double-chain structure (Matsushita et al. 2004b). Carbonation shrinkage values of ACC ranging between 0.1% and 1.0% have been reported, depending on raw materials, CO_2 concentration, and relative humidity (Matsushita et al. 2009). For real structures made of AAC, it is noticed that the degree of carbonation can reach values of 60% after 30 years (Matsushita et al. 2004a), which explains why carbonation is one of the most harmful factors affecting the durability of AAC.

5.2.4 Erosion

ACI Committee 210 (2003) defines erosion as the progressive disintegration of a solid by cavitation, abrasion, or chemical action. The committeee further explains that cavitation erosion results from the collapse of vapour bubbles formed by pressure changes within a high-velocity water flow, while abrasion erosion is caused by water-transported silt, sand, gravel, ice, or debris. Mehta and Monteiro (2006) consider the general term *surface wear* for the progressive loss of mass from a concrete surface due to abrasion, erosion, and cavitation. With abrasion, they refer to dry addition as in the case of wear on pavements and industrial floors by vehicular traffic. The term *erosion* is used to describe wear by the abrasive action of fluids containing solid particles in suspension. While the definitions of abrasion and erosion seem to be different in these two references, Mehta and Monteiro (2006) follow the same meaning of cavitation as given by ACI Committee 210 (2003).

In this book, we will consider erosion as the progressive loss of mass from a concrete surface due to cavitation or abrasion. Chemical actions will be considered separately (see Section 5.3). With abrasion, we understand the erosive action by solids, whether in dry state or in suspension. Both forms of erosion considered here, namely cavitation and abrasion, will be further explained in the following sub-sections.

5.2.4.1 Erosion by cavitation

According to Moskvin (1978), 'cavitation takes place in a fast fluid as a consequence of a decrease in pressure. Disturbance of the continuity in a stream of water is accompanied by the formation of small and then of larger vapour-filled cavities. Upon sudden collapse of these cavities as a result of a rapid change in external pressure, a large impact pressure is created. If a collapse of the cavities occurs on or near the surface of a solid body, the forces are transmitted to it, which if frequently recurring, can cause pitting of the surface material'. Two main parameters seem to influence the problem of cavitation; namely the velocity of the fluid and the roughness of the surface. A higher roughness promotes cavitation, as is the case for a higher fluid velocity.

Since the aggregate particles are typically harder than the mortar matrix, cavitation erosion initially shows a clear tendency to follow the mortar matrix, after which the aggregate particles are undermined (ACI Committee 210, 2003). Microcracks in the matrix and in the interfacial transition zone

(ITZ) contribute to cavitation damage. As soon as cavitation erosion has been initiated, the erosion rate will increase due to the increasing roughness of the surface and thus the increased promotion of cavitation. Significant damage can occur, possibly leading to complete failure of the structure as illustrated by the 1983 spillway tunnel failure at the Glen Canyon Dam in Arizona.

Increasing the concrete strength will increase the resistance against cavitation erosion. However, a strong concrete surface is not necessarily sufficient to mitigate cavitation damage, because even steel is typically damaged by this phenomenon (as clearly noticeable on pumps and marine propellers). Avoidance of cavitation erosion can be best obtained by an adequate design of the hydraulic structure, keeping pressures high and fluid velocities low. More detailed information can be found in the report of ACI Committee 210 (2003).

5.2.4.2 Erosion by abrasion

In order to clean concrete surfaces, e.g. in view of some rehabilitation activities, sand blasting is a very popular technique. Sand particles are projected onto the concrete surface, either in dry form or in suspension. Their abrasive action removes a thin layer of material including dirt particles, leaving a very clean and nice concrete surface. Abrasion is a very efficient cleaning technique when performed in a controlled way. However, the same abrasive action can severely damage concrete structures when performed in a non-controlled way: erosion by abrasion. Not only sand particles can cause abrasion. As mentioned in the definition, all solids, whether in dry state or in suspension, can contribute to abrasion. This includes traffic on roads and even pedestrians on walkways or in corridors (see Figure 5.13).

Besides roads, walkways, and stairs which are typical cases showing dry abrasion erosion, spillways, basins, sluiceways, drainage conduits, and tunnel linings are typical hydraulic structures showing abrasion by waterborne solids. In the case of hydraulic structures, abrasion erosion can be combined with cavitation erosion (see Section 5.2.4.1). In severe cases, the abrasion erosion can range in depth from a few centimetres to a few metres, depending on the flow conditions. Some remarkable examples are shown in the report of ACI Committee 210 (2003).

The abrasive erosion rate is influenced by several parameters. A first list of parameters relates to the eroding particles: size, shape, quantity, hardness, and velocity. A second list relates to the concrete quality: strength, porosity, water content, aggregate content, aggregate type, and curing. Papenfus (2003) schematically gives a very detailed list of the various influencing factors. More practically, several authors mention that a minimum compressive strength of 28 MPa (400 psi) should be considered to obtain abrasion resistant concrete surfaces (Mehta and Monteiro 2006). Liu et al. (2006)

Figure 5.13 Abrasion of concrete floor by intensive walking and sliding near a range of lockers.

conclude that the splitting tensile strength is a more effective predictive parameter than the compressive and flexural strength in determining the concrete abrasion erosion resistance to water-borne sand. Nevertheless, Neville and Brooks (2010) summarize that the resistance to abrasion is proportional to the water/cement ratio and hence to the compressive strength. They conclude that the primary basis for the selection of abrasion resistant concrete is the compressive strength.

Although strength surely is expected to be a main governing parameter concerning abrasion resistance, other aspects are also relevant. Mineral admixtures such as fly ash can enhance the abrasion resistance due to mortar densification, bleeding void reduction, and interfacial bond strength improvement. In the case of low-strength concrete, a coarser aggregate can improve the abrasion resistance. However, in high-strength concrete, the size of the coarse aggregate seems to be insignificant with respect to abrasion resistance.

5.2.5 Thermal effects

5.2.5.1 Temperature gradients

In massive concrete structures or in concrete elements made with high-performance concrete containing a high amount of cement, thermal gradients typically occur during the hardening phase. This phenomenon has been explained in detail in Section 4.4 of Chapter 4.

In hardened concrete structures, however, temperature gradients also occur due to the effect of day/night and summer/winter cycles. Deformations and deflections can be significantly influenced by thermal effects. In cases where the thermal deformations are fully or partially restrained, thermal stresses will be initiated, possibly resulting in cracking and stiffness reduction.

Bridges are a typical showcase of the importance of temperature effects on structures. Due to temperature evolutions of the environment, including the effect of solar radiation on the bridge deck, a bridge will show significant elongation when the temperature increases or shortening when the temperature drops. The daily and seasonal length variations have to be allowed for installing adequate bridge bearings and bridge joints. Malfunctioning of these features can lead to significant stresses and damage. In continuous girder bridges, temperature variations and thermal gradients, e.g. between top and bottom faces, will also induce stresses comparable to the effect of direct loading or the effect of settlements. The stresses due to thermal gradients can be calculated by means of standard design software. However, an analytical approach enabling the investigation of the effect of temperature change on serviceability of concrete structures is given and illustrated by ACI Committee 435 (1997).

For industrial buildings, temperature-induced deformations are becoming more and more relevant for sandwich façade panels consisting of an inner and outer concrete panel separated by a substantial thickness of insulation. In the case of direct solar radiation, very high temperature gradients can occur over the thickness of the façade panel, leading to highly visible deformations (outward bending) or to damage (cracking of the concrete or damage to the dowels connecting the inner and outer panels).

Besides the effect of the thermal gradient, damage to concrete structures can also be induced by the temperature level itself. Extreme temperatures, low and high, can have important effects. This is probably well-known in the case of high temperatures (fire conditions), but also holds for very cold temperatures (cryogenic conditions). Both situations will be explained in the following sections.

5.2.5.2 Cryogenic conditions

Cryogenic or very cold temperature conditions can be found in typical applications such as freeze storage rooms, –10°C to –30°C, and storage tanks for liquefied gases such as LNG, which are stored at a temperature below –162°C. Especially in view of rising energy demands after World War II and after the discovery of natural gas in the North Sea, LNG transport and storage rapidly grew in the 1970s (The Concrete Society 1981). For safety reasons, double-walled storage tanks are typically constructed for liquefied gases such as LNG. The inner or primary tank is usually made of nickel steel and the outer tank is made of reinforced or prestressed concrete. However, the inner tank itself could also be constructed of prestressed concrete (The Concrete Society 1981).

The properties of concrete in a temperature range between 20°C and cryogenic temperature depend on the moisture content. For moist or saturated concrete, the crystalline ice structure below 0°C is of importance (Waagaard 1981). At normal air pressure, the stable phase of ice is called ice I, with two variants: hexagonal ice I_h (formed between 0°C and –115°C) and cubic ice I_c (formed below –115°C). At even lower temperatures around –200°C, orthorhombic ice is formed. The density and the mechanical properties of ice depend on the crystalline ice structure, which, in turn, influences the cryogenic concrete behaviour.

Wiedemann (1982) studied the cryogenic behaviour of saturated concrete. While considering a cycle of cooling and reheating, he distinguished different temperature ranges as illustrated in Figure 5.15. Between 0°C and –20°C, the water in the larger pores freezes and becomes ice, pushing excess water to partially filled pores or to air voids. The ice, which first completely filled the larger pores will, while further cooling down, contract more than the cement matrix. Because of thermodynamic non-equilibrium, water will now move from smaller pores toward the ice in the larger pores and freeze to ice. In this temperature range, the concrete contracts,while compressive strength and ultimate strain increase with decreasing temperature because ice-filled pores are stronger than water-filled pores.

When further cooling down within the range of –20°C to –60°C, the coarser pores are totally filled with ice. This will obstruct the movement of water from the smaller pores, which will lead to stress formation in the concrete, possibly leading to cracks. The concrete will expand and the rise of the compressive strength while cooling down is partly counteracted by internal crack formation. The ultimate strain reaches a peak value.

Between –60°C and –90°C, the ice in the pores contracts more than the cement matrix, relaxing internal stresses in the concrete. Furthermore, ice filling the pores contributes to a further increase in strength while cooling down. Even in the very small pores, water is now freezing. Thanks to existing internal cracks, ample space is available and no additional stresses are initiated. The higher thermal contraction of the matrix in comparison with the aggregates, caused by a difference in coefficient of thermal dilation, results in a prestress of the interfacial transition zone (ITZ). Due to this internal prestressing effect, crack propagation starting at the ITZ is mitigated, yielding a further increase of the compressive strength of the concrete.

Further cooling between –90°C and –170°C leads to a relatively higher volume reduction of the ice in comparison with the reduction in pore volume due to contraction of the concrete. The increase in compressive strength will be less pronounced, while the ultimate strain shows a decrease. Within this temperature range, the concrete tends to become more brittle (Rostasy and Wiederman 1980).

While reheating from –170°C to –60°C, the ice, which is expanding more than the matrix, will again fill the entire pore space. Between –60°C and –20°C, expanding and melting ice in larger pores leads to an expansion of the concrete, which is higher than the expansion which was noticed during the cooling stage. Some hysteresis can thus be seen in the thermal strain curve shown in Figure 5.14. While heating from –20°C to 0°C, all ice now melts, which reduces the internal stresses and leads to a contraction of the concrete.

The significant compressive strength increase while cooling concrete to very low temperatures has been reported by many researchers (Yamane et al. 1978, Rostasy et al. 1979, The Concrete Society 1981). However, some discussion seems to exist for the compressive strength below –100°C. As an example, Waagaard (1981) reports a compressive strength increase for moist concrete from 35 MPa at 0°C to 117 MPa at –100°C; however, this is followed by a strength reduction to 110 MPa when further cooling to –156°C. The tensile strength is reported to show a similar relative evolution.

The moisture content of the concrete is of major importance while studying the influence of cryogenic conditions on the mechanical properties of concrete. The information given above was reported for moist or saturated

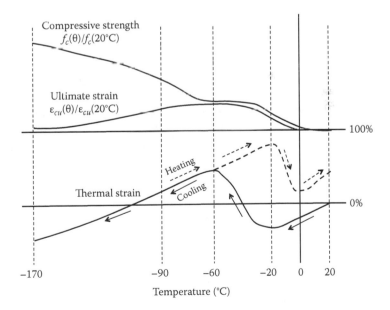

Figure 5.14 Mechanical concrete properties at cryogenic temperature, after Wiedemann (1982).

concrete. Waagaard (1981) reports a much lower strength increase (approximately 80%) for partly dry concrete and only a minor increase (approximately 20%) in the case of oven dry concrete while cooling down to –156°C. In general, compressive strength increases when reducing temperatures, the increase being larger for higher moisture contents (Browne and Bamforth 1982).

While the mechanical properties of concrete at cryogenic temperatures are much higher than at environmental temperatures, it should be mentioned that in the case of thermal cycling, water-saturated concrete can show significant losses in strength (Rostasy et al. 1979). Thermal shock loading, however, causes even more severe damage and strength reduction in comparison with thermal cycling (Yamane et al. 1978, van der Veen 1987).

Another aspect of importance for cryogenic concrete structures is the bond between reinforcing steel and concrete. It is reported that for saturated concrete, the cryogenic bond increase is relatively lower than the increase in compressive strength, while for dry concrete the bond increase is at least proportional to the compressive strength increase (van der Veen 1987). Below –170°C, slip between steel and concrete abruptly increases substantially, probably due to the initiation of internal longitudinal splitting cracks (van der Veen 1987).

In cryogenic conditions, due attention should also be given to the reinforcing and prestressing steel. Significant embrittlement upon cooling to very low temperatures should be avoided. Appropriate steel properties and qualities should be selected in order to avoid brittle failure of the entire concrete structure in cryogenic conditions. Concerning the properties of steel at very low temperatures, reference is made to literature (Sleigh 1981).

As a general finding, it can be stated that concrete is a very good material to serve in cryogenic conditions, provided that thermal cycling and thermal shocks can be avoided as much as possible. An appropriate choice of reinforcing and prestressing steel should be made in order to avoid brittle failures.

5.2.5.3 High temperature and fire

For certain structures such as industrial furnaces, e.g. in the steel industry, the concrete sometimes has to withstand very high temperatures. This is also the case in the more unfortunate conditions of fire. Major fire accidents can have a great impact on society, especially when human lives are lost. Recent fire accidents still in our collective memory include the Channel Tunnel fire in 1996 and the Mont Blanc Tunnel fire (which burned for more than two days!) in 1999.

During a fire, the high temperature causes physico-chemical changes in the concrete (Ye et al. 2007) resulting in deformations, damage, and reduction in strength and stiffness. A summary of the different phenomena on the material's level is given in the following paragraphs. Afterward, the structural behaviour during a fire is also briefly discussed.

5.2.5.3.1 Chemical and mineralogical transformations

A good indication of the chemical and mineralogical transformations of cementitious materials while heating can be obtained by differential thermal analysis (DTA) and thermogravimetric analysis (TGA), which can be simultaneously performed on the same sample. TGA measures the weight loss of the heated sample, while DTA records the temperature difference between the tested sample and an inert reference material. Figures 5.15 and 5.16 show comparative results obtained on paste of four different concrete mixes as obtained by Liu (2006). Downward peaks in the DTA curves indicate ongoing endothermic reactions. TGA curves show a gradual or accelerated mass loss. From the combination of DTA and TGA, different temperature regions can be defined with different ongoing chemical and mineralogical transformations as listed in Table 5.1. A more detailed overview can be found in literature (Annerel 2010).

Below 300°C, most traditional aggregates remain stable, leaving transformations mainly in the cement paste. At higher temperatures, significant

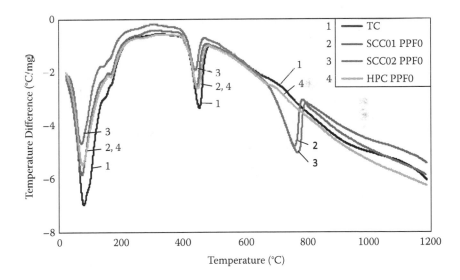

Figure 5.15 Differential thermal analysis (DTA) on paste of four different mixes (TC = traditional concrete with W/C = 0.48, SCC01 PPF0 = limestone filler based self-compacting concrete with W/C = 0.41, SCC02 PPF0 = limestone filler based self-compacting concrete with W/C = 0.48, HPC PPF0 = high performance concrete with W/C = 0.33) (Liu 2006).

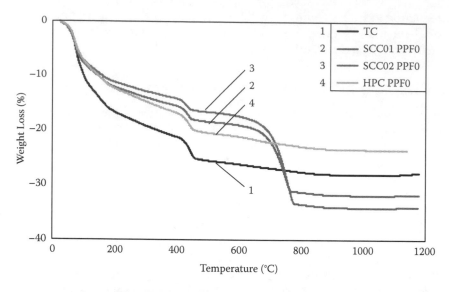

Figure 5.16 Thermogravimetric analysis (TGA) on paste of four different mixes (TC = traditional concrete with W/C = 0.48, SCC01 PPF0 = limestone filler based self-compacting concrete with W/C = 0.41, SCC02 PPF0 = limestone filler based self-compacting concrete with W/C = 0.48, HPC PPF0 = high performance concrete with W/C = 0.33) (Liu 2006).

chemical or mineralogical transformations will also occur in the aggregate particles. At 575°C, quartzite aggregate particles will show a crystal transformation from α-quartz to β-quartz, leading to a volume increase of about 5.7%. Calcareous aggregates, on the other hand, remain stable until 700°C. Above 700°C, limestone ($CaCO_3$) gets decarbonated, forming calcium oxide (CaO) and carbon dioxide (CO_2). After cooling down, the calcium oxide (resulting from the decarbonation of limestone and from the decomposition of $Ca(OH)_2$ in the cement stone) reacts with moisture from the environment, forming calcium hydroxide ($Ca(OH)_2$). This hydration process leads to a significant volume increase (44%), causing disintegration of the concrete. This explains why further concrete damage can typically occur shortly after the fire.

5.2.5.3.2 Interaction between cement matrix and aggregates

While the previous paragraphs explain some transformations and effects in the cement stone and the aggregate particles separately, the physical interaction between both phases also has to be duly considered. As mentioned before, the cement paste shrinks due to the loss of free and chemically bound water. On the other hand, the aggregate particles expand due to thermal dilation and possible mineralogical transformation at higher

Table 5.1 Transformations in cement paste with increasing temperature

Temperature range	Transformations and effects
Below 130°C	Weight loss due to evaporation of free capillary water with a maximum around 100°C. According to Liu (2006), the weight loss up to 130°C corresponds to the amount of free capillary water, while between 130°C and 1100°C, chemically bound water is released, while weight losses also occur due to decomposition of calcium hydroxide and calcium carbonate (see further).
130°C–200°C	The broad endotherm in this temperature interval is due to dehydration of various hydrated phases. Tobermorite gel and hydrated calcium sulfoaluminate are the first solid phases affected at elevated temperature (Liu 2006). Due to the loss of water, the cement paste is shrinking.
200°C–400°C	The weight loss due to loss of chemically bound water continues at a lower rate. The cement paste further shrinks.
400°C–500°C	The endotherm peak in this temperature interval is due to the decomposition of portlandite $Ca(OH)_2$. More precisely, Liu (2006) defines the weight loss between 420°C and 460°C as corresponding to the decomposition of portlandite.
500°C–700°C	Further loss of chemically bound water.
700°C–800°C	The weight loss at the temperature range from 730°C to 770°C refers to the decomposition of calcium carbonate (Liu 2006). This peak is much more prominent in the case of limestone filler-based self-compacting concrete.
Above 800°C	Further loss of chemically bound water.

temperatures. As a result of this thermal incompatibility, high stresses can be initiated within the concrete. These stresses can be partly relaxed by load-induced thermal strain (Annerel 2010).

In some cases, the interaction between cement paste and aggregates can be of a chemical nature. As an example, in dolomite aggregate, a chemical reaction can occur between magnesium phases present in the dolomite and calcium hydroxide present in the cement paste. This reaction is expansive, causing damage to the concrete.

5.2.5.3.3 Interaction between concrete and reinforcement

The thermal conductivity of concrete is substantially lower than of steel. On the one hand, this is an important advantage because even in the case of severe fire conditions only a small outer layer of concrete will be exposed to temperatures above 300°C. This is the main reason why concrete structures have a very good fire resistance in comparison with steel structures. The steel reinforcement in a structural concrete element is somewhat protected from the heat source by the concrete cover. On the other hand, the steel reinforcement cage can heat much faster than the concrete in the core, due

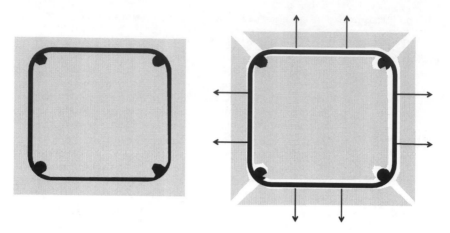

Figure 5.17 Thermal incompatibility between concrete and steel cage due to faster heating of the steel.

to the much higher thermal conductivity. This can lead to another problem of thermal incompatibility, now between the concrete and the steel cage. Due to its higher temperature, the steel cage wants to expand more than the concrete in the core. As a result, as illustrated in Figure 5.17 for the case of a reinforced concrete column exposed to fire conditions, the concrete cover can be pushed off over a larger area. Figure 5.18 shows the example of a real structure suffering from this phenomenon.

5.2.5.3.4 Explosive spalling

A more important problem, especially in the case of high-strength concrete, is the occurrence of explosive spalling when exposed to rapid heating. Due to the very low permeability of high-strength concrete, water vapour cannot easily escape from the pores. As a result, a pore pressure builds up in the cement paste. Upon further heating, the pore pressure can reach high levels, causing important internal stresses near the concrete cover. This can result in sudden explosive spalling. According to Khoury (2000), explosive spalling generally occurs under the combined action of pore pressure, compression in the exposed surface region (induced by thermal stresses and external loading), and internal cracking. Due to the mutual interaction between pore pressure and internal cracking, ongoing research focuses on the coupling of a fracture mechanics model and a pore pressure approach in order to better understand and predict fire spalling behaviour (Lottman et al. 2011).

Spalling can be limited to the pushing off of small pieces of the concrete cover or the formation of small craters in the surface. Nevertheless, in some cases fire spalling can lead to structural failure because the loss of larger concrete layers can severely reduce the structural load-bearing capacity.

Figure 5.18 Concrete cover pushed off at larger areas due to thermal incompatibility between concrete and steel cage.

In laboratory conditions, unreinforced cementitious specimens can totally disintegrate due to explosive spalling while rapidly heating in a furnace (Liu 2006).

Many parameters have an influence on the spalling risk of concrete elements exposed to fire (Boström and Jansson 2007). Major influence is noticed by the pore structure and the moisture content of the concrete. However, further parameters are linked to element size and geometry, heating rate, and pre-loading conditions.

Normal strength concrete does not typically show spalling behaviour when exposed to rapid heating. Although the spalling risk is more pronounced for high-strength concrete, it can also occur in the case of self-compacting concrete (Ye et al. 2007, Boström and Jansson 2007, Fares et at. 2011) or, in general, in any type of concrete with a more dense pore structure.

The addition of polypropylene fibres (PP) is widely used as an effective method to prevent explosive spalling. Upon the melting of the PP fibres, no significant increase in total pore volume is obtained, as is sometimes thought. However, the connectivity of the pores increases due to melting

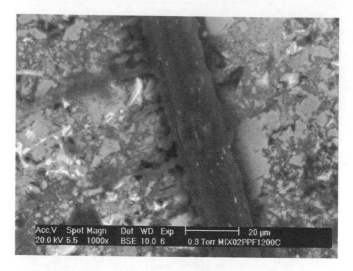

Figure 5.19 Concrete specimen with PP fibre heated up to 200°C.

of the fibres, leading to an increase in permeability (Liu et al. 2008). An electron microscope image of a melting PP fibre is shown in Figure 5.19, taken from a concrete specimen heated at 200°C. When the PP fibres melt, the melted fibres will be absorbed by the paste through the pores. After melting, the remaining fibres (the fibre channels) and the pores connect with each other and form a better connected pore network. In this way, the pressure build-up is reduced, as well as the explosive spalling risk.

5.2.5.3.5 Mechanical properties at high temperature

Due to the previously mentioned chemical and mineralogical transformations as well as the interaction between matrix and aggregates, the mechanical properties of concrete are negatively affected by exposure to high temperatures. Normal strength concrete will typically show a 10% to 20% strength reduction at a temperature of 300°C, while a 60% to 75% strength reduction is noticed at 600°C. The Young's modulus evolves in a similar way. For high strength concrete, a strength loss up to 40% can be noticed at temperatures below 450°C (Phan and Carino 2000).

Several parameters influence the residual compressive strength after fire exposure. Of course, the duration and the intensity of the heating cycle are of main importance, as well as the loading level during the heating cycle. Furthermore, the storage conditions after the heating cycle are important. Typically, after a fire and the extinguishing actions of the fire brigade, the damaged structure is exposed to weather conditions before repairing. According to Annerel (2010), during the first two months after cooling, the presence of water will result in additional strength loss due to the hydration of free lime, with a volume increase as a result. On the other hand, some chemical bonds might be restored due to rehydration of gel and unhydrated cement particles, which could lead to the partial recovery of strength compared to the strength at high temperature.

It is, however, very difficult to predict in a general way the residual strength of fire-damaged concrete structures. A detailed damage assessment will have to be performed case by case. This can consist of non-destructive strength evaluation on site (e.g. rebound hammer) or laboratory tests on drilled cores. New diagnosis techniques based on colour changes of the cement paste during fire have been proposed recently by Annerel (2010).

5.2.5.3.6 Structural behaviour

When a structure undergoes temperature changes, deformations will be induced. As long as the imposed deformations can occur freely, no stresses will be induced. However, in general, the thermal deformations will typically be restrained in two different ways:

- Internal restraint: When a temperature gradient exists over a cross section, strains have to remain compatible. This will lead to partial restraint of the free deformation of adjacent fibres of a section called internal restraint. This is similar to what happens in the case of thermal stresses in hardening massive elements due to the heat of hydration (see Section 4.4 of Chapter 4).
- External restraint: The thermal deformation of structural elements undergoing a temperature change will be (partially) restrained by adjacent elements or other structures which are not exposed to the same change.

Due to the (partial) restraint of thermal deformation, either internal or external restraints or a combination of both, stresses will be induced in the structure. This is also valid in case of fire-exposed structures or structural elements. Due to a high temperature increase, stresses can become very high. Consider as an example a temperature increase of 100°C and a coefficient of thermal expansion of $10^{-5}/°C$. When a concrete element having a Young's modulus of 37000 N/mm² is fully restrained, a stress of 37 N/mm² results. Although this stress level is high enough to cause damage, it has to be well interpreted considering a few other phenomena:

- Upon heating a loaded structure, an additional strain called transient strain, is introduced. This transient strain will help relax the stress levels (Annerel 2010).

- At higher temperatures, the mechanical properties of the concrete will be reduced, as explained before. This also holds for the Young's modulus. As the stiffness is reduced, the restrained stresses will also be reduced. However, it is also clear that at the same time, the concrete strength is reduced.

A detailed study of the load-bearing capacity of structures or structural elements under fire conditions is out of the scope of this textbook. Second-order effects might also be induced, e.g. reducing the buckling load of columns under fire conditions. Reference is made to relevant standards and models in order to check the fire resistance of concrete structures.

5.2.6 Crystallisation and discolouring due to moisture movement

Moisture movements in concrete elements due to inadequate water tightness of coatings or roofing can lead to leaching. Calcium ions are dissolved in the moving water and transported toward the concrete surface. After evaporation of the water and reaction with carbon dioxide from the environment, whitish calcium deposits or efflorescence can build up on the concrete surface. As an example, Figure 5.20 shows some stalactites formed on the bottom of a concrete slab in a cellar, while Figure 5.21 shows some calcium deposits on a vertical surface. When other minerals present in the concrete are dissolved, the deposits can have other colours such as yellow or brown.

Depending on wetting and drying cycles and on the local super saturation of the water inside the concrete element, deposits can also form inside

Figure 5.20 Formation of stalactites.

Figure 5.21 Formation of calcium deposits.

the concrete along micro cracks. On the one hand, the internal crystallisa-tion process can lead to a weakening of the structural element, as in the case of Figure 5.22 where internal crystallisation weakened some concrete balconies near the Belgian coast. On the other hand, internal crystallisation in micro cracks can cause some self-healing, restoring to some extent the 'water tightness' of the concrete.

In many cases, water movement on concrete structures can also lead to vegetation or moss formation or to zones with different levels of gray (darker and lighter zones) depending on the way dirt particles are washed away, e.g. by the rain. This is only an esthetical problem without serious harm to the concrete structure.

5.2.7 Pop-outs

Pop-outs can occur due to reactive aggregate particles in the concrete. When the aggregate particle is positioned near the concrete surface, the expansion can lead to pop-outs or small craters in the concrete surface. Reference is made to Chapter 3, Section 3.4, where the effect of expansive aggregate particles or other inclusions is described in more detail.

Figure 5.22 Internal crystallisation in balconies. The picture shows the crystallisation layers immediately after breaking the balcony. In some cases, several parallel crystallisation layers can be noticed, which could be referred to as "puff pastry concrete" (courtesy of SCICON Worldwide bvba).

5.3 CHEMICAL ACTIONS

5.3.1 Alkali silica reaction (ASR)

Alkali aggregate reaction (AAR) is a heterogeneous chemical reaction between a chemically instable solid phase present in aggregate particles, and an alkali rich liquid phase formed by the pore solution present in the concrete pores. The pore solution has a high alkalinity with a pH value of 12.5 or higher. In the presence of alkalis (sodium, Na, and potassium, K) in the concrete, the pore solution mainly contains dissolved alkali hydroxides (K^+OH^- and Na^+OH^-) and minor quantities of other ions (Ca^{2+}, SO_4^{2-}). In this environment, the chemically unstable aggregate phases show a reaction, possibly resulting in internal swelling and damage to the concrete. On the concrete surface, alkali aggregate reaction yields a typical and extensive map cracking, as illustrated in Figure 5.23. A gel can often be found in the cracks and, in advanced cases, spalling of surface concrete can occur.

Depending on the unstable aggregate phases, two main types of alkali aggregate reaction can be defined: alkali silica reaction (ASR) and alkali carbonate reaction (ACR). While ACR is less common in practice, and thus causes considerably fewer damage cases, ASR is very relevant in many parts of the world and results in a high number of severely attacked structures and important economical losses (Swamy 1992).

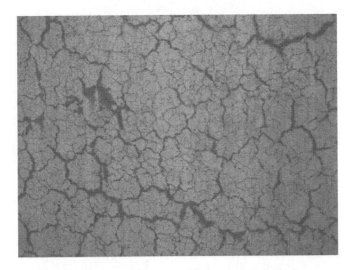

Figure 5.23 Extensive map cracking on the concrete surface due to alkali aggregate reaction.

The first cases of alkali aggregate reaction were observed in the 1920s and 1930s in California USA, showing severe cracking only a few years after construction. It was Stanton (1940) who first demonstrated the existence of a deleterious reaction between the constituents of the concrete. Nevertheless, it took many more years before the mechanism of alkali aggregate reaction became clearer. A first international conference on alkali aggregate reaction was organized in 1975 in Reykjavik (Asgeirsson 1975), with additional conferences following later in the 1970s and 1980s. In many countries, the first cases of alkali silica reaction were diagnosed in the 1980s, e.g. in Belgium in the Kontich Bridge around 1985 (De Ceukelaire 1986, 1988, 1991). Gradually, the concrete industry became well aware of the risk of alkali silica reaction and appropriate preventive measures have been defined. Nevertheless, alkali silica reaction in concrete structures is still an important issue in many parts of the world.

5.3.1.1 Mechanism

5.3.1.1.1 Alkali silica reaction

It is generally accepted that alkali silica reaction is linked to the presence of silicon dioxide, SiO_2, in the aggregates, and more specifically whether it is crystalline or amorphous. Most aggregates contain silicon dioxide, often in the crystalline form of mineral quartz, although other minerals can also occur. Due to its chemically stable bonds, quartz is quite unreactive and it will not be affected by acids or alkalis. Unfortunately, silicon dioxide can also be present in a disordered, amorphous form. These poorly crystallised silicas are more prone to chemical reaction. To a certain degree, silica can be replaced by water, yielding an amorphous hydrous silica which can be highly reactive in an alkali rich environment. Although this principle is clear and well accepted, the detailed reaction process is probably not yet fully understood because of its extreme complexity. Some basic principles of the reaction process are given hereafter.

The chemistry of silica dissolution, which results in the formation of alkali silicate solutions, is very complex (Helmuth et al. 1993). However, it is only by understanding the dissolution of soluble silica and the formation of swelling alkali silica gels that a good understanding of the ASR damage mechanism can be reached. This not only involves the role of alkalis (Na and K), but also the role of calcium ions.

Probably the first comprehensive attempt to fundamentally explain the mechanism of ASR damage was made by Powers and Steinour (1955, Parts 1 and 2). The strength of their approach was the combination of chemical aspects with expansion studies, enabling conclusions in terms of 'safe' and 'unsafe' reactions. Based on their studies of opal rock as aggregate particles, Powers and Steinour concluded that the relative amounts of calcium and alkalis in the reaction product determine whether swelling damage will occur. High calcium content alkali silica gels will not cause swelling damage. In case of a lack of calcium ions, low calcium content gels are formed, which produce considerable swelling and possible damage to the concrete. In order to have a 'safe' gel formation, Powers and Steinour also concluded that silica must be able to diffuse out from the gel, while water, calcium, and alkalis must be able to diffuse into the gel. Figure 5.24 shows the principles of the model developed by Powers and Steinour (1955). It is redrawn here for historical reasons, while it is clear nowadays that this model does not duly take into consideration some important aspects, e.g. the fact that in the presence of alkali, very little calcium will be present in the pore solution because the high pH value will substantially decrease the solubility of $Ca(OH)_2$. Nevertheless, the pioneering research by Powers and Steinour was certainly remarkable.

A closer look at the pore solution of concrete reveals that, because of the presence of alkalis in cement or in other constituent materials, it contains alkali ions (Na^+ and K^+) and hydroxyl ions (OH^-). In principle, the $Ca(OH)_2$ resulting from cement hydration could also dissolve in the pore solution. However, due to the high alkali content, the calcium is nearly insoluble. The pore solution can thus be considered as a fluid with high concentrations of hydroxyl ions balanced by sodium and potassium ions. It is now accepted that the hydroxyl ions initiate the chemical reaction of the silica, while the alkali ions are only relevant when they are incorporated into the gel (Swamy 1992, Bazant and Steffens 2000).

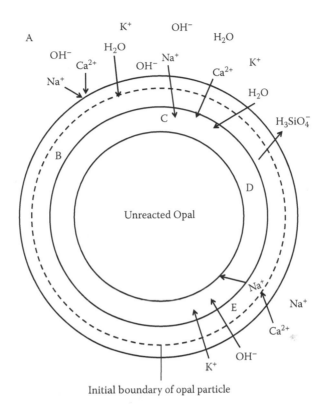

Figure 5.24 Principles of the ASR damage model by Powers and Steinour (1955). A: Pore
solution containing H_2O, K^+, Na^+, Ca^{2+}, OH^- and $H_3SiO_4^-$; B: initial non-
swelling C-N-S-H gel, with calcium content depending on the sodium concentration; C: diffusion through gel to react with opal, expansive swelling
N-S-H gel in case of low Ca^{2+} content, safe non-swelling C-N-S-H gel in
case of high Ca^{2+} content; D: diffusion of $H_3SiO_4^-$ out of the gel in case of
safe reaction, while water, Ca^{2+} and Na^+ diffuse into the gel; E: regeneration
of Na^+ when Ca^+ reacts with gel.

The chemical reaction between the alkali pore solution in the concrete
and the amorphous silica in the aggregates is commonly schematized by
following equations (Swamy 1992, Soutsos 2010):

$$4SiO_2 + 2NaOH \rightarrow Na_2Si_4O_9 + H_2O \qquad (5.1)$$

$$SiO_2 + 2NaOH \rightarrow Na_2Si_3 + H_2O \qquad (5.2)$$

The alkali-silica gel which is produced is hydrophilic, resulting in water
absorption and expansion. The given chemical equations, however, are
simplified, because the real chemical composition of alkali-silica gel is

volatile and not exactly known. While Powers and Steinour (1955) gave a fundamental explanation of the complex reaction schemes, they more or less restricted ASR to topochemical processes. It is now accepted that ionic reactions within the pore solution are also of great importance to ASR. A multi-stage process is typically considered now to explain the mechanism of expansive ASR (Bazant and Steffens 2000).

The first stage of the reaction scheme is the hydrolysis of the amorphous silica by the hydroxyl ions, resulting in the formation of a gel-like layer on the surface of the aggregates. For highly alkaline solutions, the monomer $H_2SiO_4^{2-}$ is formed, while a somewhat lower pH (toward 11.2) leads to the monomer $H_3SiO_4^-$.

In the second stage, the negatively charged gel species attract positive ions from the pore solution. Depending on the type of cations, swelling will occur or not. In a calcium rich pore solution, Ca^{2+} will be attracted by the gel, resulting in the formation of calcium silicate hydrate, C-S-H, thus transforming the gel into a rigid structure. This process is similar to puzzolanic reaction schemes and will not cause swelling damage. When the pore solution is poor in calcium, mainly the alkali ions (Na^+ and K^+) are attracted by the gel. A viscous and expanding gel is produced, potentially causing damage to the concrete.

Ichikawa and Miura (2007) further detail the mechanism involved in swelling damage of ASR, as illustrated in Figure 5.25. Within their modified ASR model, the first stages are similar to what has been previously mentioned: hydrolysis of the amorphous silicate and subsequent attraction of alkali ions into the gel, forming a hydrated alkali silica gel. But in order to cause swelling damage, further steps are required, according to Ichikawa and Miura (2007). Due to the reaction of OH^- with the amorphous silica and the uptake of alkali ions by the gel, the solubility of $Ca(OH)_2$ will increase, leading to an increased content of Ca^{2+} ions in the pore solution. As a result, Ca^{2+} ions penetrate into the soft alkali silica gel, forming a rigid rim of C-S-H around the reactive aggregate particle. This rim, however, still allows the diffusion of alkali and hydroxyl ions to the inner soft gel, making it further expand. Inside the aggregate, a highly expansive pressure is now developed, and cracking will occur when the pressure exceeds the strength of the aggregate surrounded by the reaction rim and the cement paste.

It is further stated by Ichikawa and Miura (2007) that ASR will not result in the deterioration of concrete when the formation of the alkali silica gel is completed before the formation of the rigid reaction rim around the aggregates. As a result, reactive but tiny silica-rich aggregates (e.g. fly ash) will not induce damage because they are entirely converted to alkali silicate before the formation of the reaction rim. With this conclusion, the author show again the resemblance of puzzolanic reactions and the alkali silica reaction.

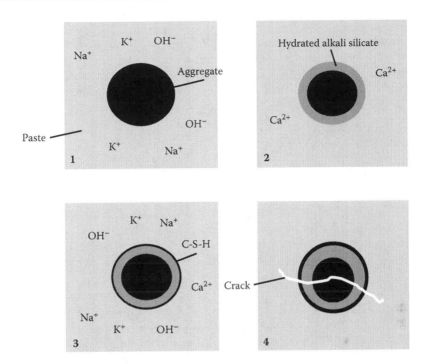

Figure 5.25 Modified model of ASR (Ichikawa and Miura 2007).

The effect of tiny aggregate particles on ASR expansion has also been experimentally confirmed by Multon et al. (2010), although they do not refer to the concept of rigid rim as conjectured by Ichikawa and Miura (2007). Multon et al. (2010) report some experimental results showing that no expansion was measured on mortars containing small reactive particles (<80 μm), while the largest expansion was obtained with coarser particles (with particle size ranging from 0.063 mm to 1.25 mm). They explain this behaviour by means of a phenomenological model, assuming migration of the alkali silica gel into the pore structure of the matrix surrounding the reactive aggregate along a distance equal to the aggregate size. Although the same conclusions concerning the effect of tiny aggregate sizes are obtained as within the rigid rim concept, it illustrates that a lot of debate still exists concerning the real mechanisms behind the not yet fully understood swelling damage due to ASR. Concerning the effect of particle size on ASR expansion, it is also important to mention that for particle sizes larger than 0.15 mm, Hobbs (1988) reports a decreasing expansion with increasing particle size. It thus seems that the behaviour at the micrometre level is different from the behaviour at the millimetre level.

5.3.1.1.2 Alkali carbonate reaction

The alkali carbonate reaction is less common, although several cases have been reported. Some dolomitic aggregates containing $CaMg(CO_3)_2$ can react with hydroxyl and alkali ions according to a complex reaction mechanism, which is not yet fully understood. Two main subsequent reactions can be considered in the case of ACR:

$$CaMg(CO_3)_2 + 2NaOH \rightarrow Mg(OH)_2 + CaCO_3 + Na_2CO_3 \qquad (5.3)$$

$$Ca(OH)_2 + Na_2CO_3 \rightarrow CaCO_3 + 2NaOH \qquad (5.4)$$

The first reaction is the so-called dedolomitization, transforming dolomite ($CaMg(CO_3)_2$) into brucite ($Mg(OH)_2$) and calcite ($CaCO_3$). This process is not expansive, so no expansion cracks will be formed. On the contrary, a reduction of 5.1% of the total solid volume is obtained.

The second reaction, a dissolution–precipitation process, describes the formation of secondary calcite, which is deposited in the voids of the interfacial transition zone (ITZ) surrounding the dolomite aggregate. Although the absolute volume increases with 10.2%, no expansion stresses are obtained as the products are formed in the voids of the ITZ. The given simplified reaction shows that Na^+ ions are regenerated through the water-soluble NaOH, maintaining a high alkalinity, favouring a potential alkali silica reaction.

The reaction process further proceeds near the dolomite surface forming a narrow rim of hydrotalcite ($6MgO \cdot Al_2O_3 \cdot CO_2 \cdot 12H_2O$) and additional calcite, while increasing the porosity of the dolomite.

As the cause of this damage process is typically ascribed to the dedolomitization (reaction of dolomite with hydroxyl ions and subsequent crystallisation of brucite), it is commonly considered to be a different type of alkali aggregate reaction in comparison with alkali silica reaction. However, the damage process due to the alkali carbonate reaction (ACR) is controversial.

In the 1990s, Katayama stated that ACR, in fact, was a combination of expansive ASR of amorphous silica and harmless dedolomitization of dolomitic aggregate (Katayama 1992). By means of petrographic analysis in combination with SEM-EDS analysis, Katayama (2010) recently was able to give more evidence for this statement. Dolomitic aggregates did not develop expansion cracks unless ASR was involved. It is the alkali silica gel which is responsible for the crack formation in concrete. This is further confirmed by Grattan-Bellew et al. (2010). For ACR-susceptible aggregates, Grattan-Bellew et al. found a good correlation between the amount of quartz in the aggregates and the expansion of the concrete prisms, referring to the more important role of ASR. They finally conclude that the alkali carbonate reaction is just a variant of the alkali silica reaction, or in short ACR = ASR.

It should be mentioned that the dolomitic aggregate studied by Katayama contained a substantial amount of quartz. In the case of a relatively pure dolomitic aggregate with a very low quartz content, dedolomitization occurs without the expansive formation of ASR gel. Due to this dedolomitization, the porosity of the dolomite increases and the mechanical properties can be reduced. This process will not be visible from the outside of the concrete element because it does not lead to expansive crack formation. Consequently, ACR is typically considered to be harmless, neglecting potential strength reduction due to the dedolomitization. Visible damage will only occur in the case of simultaneous ASR. For this reason, in the following paragraphs, the focus will be on ASR.

5.3.1.2 Influencing parameters

For ASR to occur, three necessary conditions have to be fulfilled:

- Potentially reactive aggregates have to be present in the concrete.
- The alkali concentration in the pore solution has to be sufficiently high.
- Moisture has to be present in sufficient quantities.

In this respect, reference is typically made to so-called pessimum concentrations. This means that the three required elements (reactive silica, alkalis, and water) have to be present in unfavourable quantities, meaning quantities which are most beneficial for the reaction to proceed. As the consequences of the reaction are not positive, the term pessimum is used instead of optimum.

The reactivity of the aggregate particles is dependent on several parameters including the content of amorphous silica in the particles, the fraction of reactive particles, and the particle size. Typically, finer aggregate particles show a more intense reactivity, although in the case of very fine particles the expansion can be very limited.

The alkali concentration in the pore solution typically depends on the alkali content of the cement, although alkalis can also be provided to the system through admixtures, additives, mixing water, and even the aggregates. Alkalis can also enter the concrete from the environment when exposed to sea water, de-icing salts, or industrial solutions. The term alkali refers to the alkali metals sodium and potassium for which the combined concentration in concrete is typically expressed in terms of the sodium oxide equivalent, according to the following expression:

$$Na_2O_{eq} = Na_2O + 0.685\ K_2O \qquad\qquad (5.5)$$

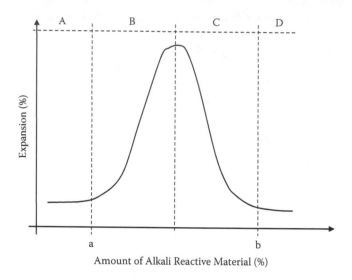

Figure 5.26 Pessimum concentration.

The moisture content depends on the environmental conditions and on the dimensions of the element. Larger elements typically maintain a higher internal relative humidity, even in the case of a dryer environment, while more slender elements will more rapidly show a dryer internal relative humidity in a drying environment.

The concept of pessimum concentration is illustrated in Figure 5.26, which shows the expansion due to ASR as a function of the concentration of alkali reactive material. The expansion reaches a maximum level for a certain concentration of alkali reactive material, called the pessimum concentration. The levels a and b are the pessimum boundaries. When the concentration of alkali reactive material is outside the interval [a,b], ASR can still occur but very little destructive expansion will occur. This is the case in the designated zones A and D. In zone A, it is clear that expansion will not occur due to a lack of alkali reactive material. In Zone D, ASR occurs simultaneously in larger areas of the concrete element, forming a widely spread quantity of alkali silica gel. As a consequence, the alkali concentration of the pore solution drops very quickly and the gel further reacts with calcium ions, forming stable C-S-H which further densifies the pore structure. In zones B and C, reaction will occur as well as expansion, resulting in an increased risk of cracking.

The pessimum boundaries, defining the concentration levels at which ASR can be very destructive, are not clearly defined and depend on several parameters including cement type, concrete composition, and temperature. In the case of high performance concrete, ASR expansion seems also to depend significantly on the air content (Ferraris 1995).

The permeability of the concrete, as mainly governed by the water/cement ratio, also has an important influence. On the one hand, the reaction rate of ASR depends on the mobility of ions and water, which in turn depends on the permeability of the concrete. A higher permeability might thus increase the reaction rate. On the other hand, the expansive pressures caused by the gel formation can be reduced in case of absorption in a higher pore volume. This means that in case ASR occurs, pressures will be lower in a more porous system. While this statement would be supported by the theory of Multon et al. (2010), it is not supported by the theory of Ichikawa and Miura (2007). This again shows the debate that still exists due to the very complex nature of ASR.

Supplementary cementitious materials (SCM) also seem to have a significant influence on expansion due to ASR. While SCMs typically contain alkali themselves, the main mechanism by which SCMs reduce expansion due to ASR is by lowering the alkali concentration in the pore solution of the concrete, probably due to the presence of alumina in the SCM (Thomas 2011). With higher replacement levels, a higher reduction in the concentration of alkali-hydroxides is obtained in the pore solution. Silica fume is very efficient in reducing the hydroxyl ion concentration, although a slow increase is noticed again after three months. This increase is not noticed in the case of other SCMs. Metakaolin, low-calcium fly ash, and slag also efficiently reduce the hydroxyl ion concentration. High calcium fly ash and high alkali fly ash show a lower efficiency and require higher dosages to reduce the pore solution alkalinity (Thomas 2011). Following the results of Shehata and Thomas (2000, 2002), fly ashes with low alkali and calcium content ($Na_2O_{eq} < 1\%$, $CaO < 20\%$) are generally effective in controlling ASR expansion, considering 25% cement replacement. For higher calcium content ($CaO > 20\%$), ASR expansion increases linearly with increasing calcium content. In the case of fly ash with a high alkali content ($Na_2O_{eq} > 5\%$), ASR expansion cannot be effectively controlled at a 25% replacement level, irrespective of the calcium content.

The influence of chemical admixtures (plasticizers, retarders, accelerators) on ASR primarily seems to depend on their alkali content. Chemical admixtures with a high alkali content typically seem to increase the risk of ASR expansion. This is related to the overall alkali content in the concrete.

5.3.1.3 Mitigation

In order to mitigate ASR, actions can be taken to make sure that at least one of the three required conditions for ASR is not fulfilled. The most straightforward measure, of course, is to *avoid potentially reactive aggregates*. However, quite often this is neither practical nor economical because it depends on the availability of suitable local materials. Care should also be taken in examining the available materials. Although the application of limestone aggregates might sound like a sure and safe action; surprisingly, some limestone is potentially reactive. This is the case for limestone from the region of Tournai, Belgium, which is a siliceous limestone containing amorphous or poorly crystalline silica.

A more practical measure is the *limitation of the alkali content of the concrete*. Typically, this is achieved by prescribing a cement with a low alkali content (a so-called LA cement), although it is clear that other constituents can also contain alkali. A safer approach is to check the total alkali content of the concrete by considering the alkali content in all constituent materials. In many countries, the total alkali content (expressed as Na_2O_{eq}) of the concrete is required to remain below 3 kg/m³. This threshold may be increased in case of the *application of supplementary cementitious materials* (SCM), because they typically reduce the alkali content of the pore solution in spite of containing alkali themselves.

Protecting the concrete from rain and other sources of water, or in other words *keeping the concrete dry*, would also help prevent expansion due to ASR. However, this is not a realistic approach in 'wet countries' such as Belgium. Sealing the concrete with protective coatings does not seem to be efficient because the internal moisture in the concrete element might already be sufficient to initiate ASR. Nevertheless, coatings could reduce the expansion rate. Reducing the water/cement ratio of the concrete will reduce the permeability, and thus obstruct the water supply to the reactive aggregates. The resulting effect, however, is a mere slowing down of the expansive reaction.

Mitigation of ASR by *special chemical admixtures* could also be an option. Lithium and barium salts seem to show promising results toward controlling ASR expansion (Feng et al. 2010, Bulteel et al. 2010). Their preventive action is related to the formation of non-swelling lithium and barium silicate hydrates. The required dosage depends on the alkali content of the concrete, the properties of the potentially reactive aggregate, and the type of lithium or barium salt. However, application of these special chemical admixtures remains very limited until now, probably because of the high cost. Lithium treatment of ASR-affected existing structures in order to stop ongoing ASR has also been investigated (Ueda et al. 2011), but this is beyond the scope of this textbook.

5.3.1.4 Example

Cases of ASR can easily be found worldwide. An interesting example is the Montreal Olympic Stadium, which was built for the Olympic Games of 1976. However, despite the fact that ASR is better known nowadays, it can still be found in more recent structures. An example is given in Figure 5.27, which shows concrete columns within a frame supporting a crane bridge in open air. Severe damage due to ASR occurred within less than five years after construction. At the top of the columns, concrete is spalled, yielding

Figure 5.27 ASR in concrete columns.

a risk of injuries for the technicians operating the bridge. The concrete contains limestone aggregates from Tournai in Belgium and is made with ordinary Portland cement. In spite of the good concrete quality, severe ASR expansion rapidly occurred in the columns in open air, which were exposed to rain. Identical columns positioned in the construction hall protected from rain also show damage caused by ASR, although it is much less pronounced. This illustrates the important role of moisture on the ASR expansion rate.

5.3.2 Sulfate attack and delayed ettringite formation

According to Skalny et al. (2002), sulfate attack is the term used to describe a series of chemical reactions between sulfate ions and the components of hardened concrete, principally the cement paste, which is caused by the exposure of concrete to sulfates and moisture. This definition includes a complex set of overlapping chemical reactions, which are duly influenced by environmental and physical conditions.

Ground water and sea water typically contain sulfates. High sulfate concentrations may also be found in industrial wastewater. Sulfate ions will react with the hydration products present in cementitious materials, causing *chemical sulfate attack*. In partly submerged elements, sulfate salts can crystallise on the drying surface. The crystallisation pressure can damage the concrete surface, causing so-called *physical sulfate attack*. In cases of accelerated curing conditions involving high temperature, a special case of chemical sulfate attack can occur at later age. This is called *delayed ettringite formation*. These degradation mechanisms, all linked to the presence of sulfate, will be explained in some more detail in the following sections.

It should be mentioned, however, that sulfate attack on concrete is a very complex and sometimes even confusing item (Neville 2004). The classical distinction between chemical and physical sulfate attack (which will also be adopted in this book for reasons of clarity) is not entirely justified scientifically because 'physical' sulfate attack clearly also involves 'chemical' processes. Furthermore, processes of sulfate attack quite often run parallel with other degradation processes. When concrete is exposed to $NaSO_4$, it can be degraded by both sulfate attack and alkali silica reaction (ASR, see Section 5.3.1). Sulfuric acid will lead to a combination of acid attack and sulfate attack. When microorganisms are involved, as in sewer pipes, it will become an even more complicated degradation mechanism; this is called biogenic sulfuric acid attack (see Section 5.3.4).

In literature, a distinction is often made between internal and external sulfate attacks. *Internal sulfate attack* occurs when the source of sulfate is within the concrete, e.g. caused by cement with a too high sulfate content or by sulfate-containing aggregates. This has already been mentioned in Chapter 3. Delayed ettringite formation is a special case of internal sulfate attack, as will be explained further on. *External sulfate attack* refers to the situation where the sulfate comes from an external source, e.g. from sea water, wastewater, or ground water.

Sulfate attack can lead to a variety of damage patterns, such as spalling, delamination, cracking, and loss of cohesion. Skalny et al. (2002) give a comprehensive list of complex physico-chemical processes which may be involved in a typical sulfate attack:

- Dissolution or removal from the cement paste of calcium hydroxide
- Complex and continuous changes in the ionic composition of the pore liquid phase
- Adsorption or chemisorption of ionic components present in the pore liquid phase on the surface of the hydrated solids present in the cementing system
- Decomposition of still unhydrated clinker components

- Decomposition of previously formed hydration products
- Formation of gypsum
- Formation of ettringite
- Formation of thaumasite
- Formation of brucite and magnesium silicate hydrate
- Formation of hydrous silica (silica gel)
- Penetration into concrete of sulfate anions and subsequent formation and repeated recrystallisation of sulfate salts

Most of these processes will be explained in the following sections. However, it is not the intention of this general textbook to provide advanced chemical details of concrete damage. Further reference will be made to literature.

It should be mentioned that in some cases of sulfate attack (e.g. ettringite formation), expansion of the cement paste causes concrete damage; however, in other cases, it concerns a decomposition of hydration phases. In some cases, it is debated whether a reaction causes expansion, e.g. in the case of gypsum formation. In cementitious systems, most of the chemical reactions involving sulfate ions show some chemical shrinkage (Skalny et al. 2002), meaning that the sum of the absolute volumes of the reacting products is larger than the absolute volume of the reaction products. Nevertheless, in the case of a reaction showing chemical shrinkage, an expansion of the cement paste can occur. Several theories have been developed to explain the resulting expansion. A comprehensive overview is reported by Skalny et al. (2002). Further details on volumetric relations and expansions in case of sulfate attack can also be found in Clifton and Ponnersheim (1994).

5.3.2.1 Chemical sulfate attack

5.3.2.1.1 Mechanism

The mechanism of a chemical sulfate attack depends on the type of ions that are combined with the sulfate ions. In a natural environment, sulfate ions are most commonly combined with alkali or calcium ions, although magnesium ions may also be significantly present in some cases. An overview of the main chemical reactions is given below.

5.3.2.1.1.1 SODIUM SULFATE (Na$_2$SO$_4$)

When a Portland cement-based system is attacked by a sodium sulfate solution, a set of chemical reactions takes place as summarized by Skalny et al. (2002) in Figure 5.28. In a simplified way, it can be summarized that SO$_4^{2-}$ ions enter the pore solution and react with Ca^{2+} ions provided by

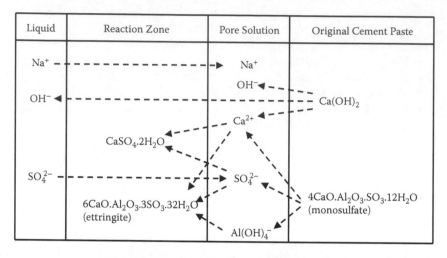

Figure 5.28 Overview of reactions for a Portland cement-based system in contact with a sodium sulfate solution (after Skalny et al. 2002).

the dissolution of Portlandite ($Ca(OH)_2$), forming gypsum ($CaSO_4 \cdot 2H_2O$). Furthermore, monosulfate, formed during hydration of Portland cement, will be transformed to ettringite.

The formation of ettringite is an expansive process, causing stresses and cracks in the cement matrix. At high sulfate concentrations, gypsum is the main reaction product because ettringite will become unstable at a pH value below about 11.5 (Neville 2004). In that case, after consumption of the Ca^{2+} ions provided by dissolution of Portlandite, the C-S-H hydration phase will be decomposed, reducing the C/S ratio. Due to this decalcification of C-S-H, a gradual loss in strength is obtained.

In addition to the chemical reactions involving sulfate ions, the sodium ions will also enter the pore solution, leading to a potential risk of ASR (see Section 5.3.1). Similar reactions can be found in the case of other alkali sulfate solutions such as potassium sulfate (K_2SO_4).

5.3.2.1.1.2 CALCIUM SULFATE ($CaSO_4$)

When exposed to calcium sulfate, ettringite formation due to transformation of monosulfate is the main damage mechanism, due to its expansive nature.

$$4CaO \cdot Al_2O_3 \cdot SO_3 \cdot 12H_2O + 2Ca^{2+} + SO_4^{2-} + 24H_2O \rightarrow$$
$$6CaO \cdot Al_2O_3 \cdot SO_3 \cdot 32H_2O$$

(5.6)

As the attacking sulfate solution also provides the Ca^{2+} ions, no decalcification of C-S-H and no associated strength loss occur in this case. Initially, the concrete strength will even increase due to the formation of ettringite filling the pores. However, upon further reaction and ettringite formation, expansive stresses will occur, causing cracking of the cement matrix and thus reducing the concrete strength at this stage.

5.3.2.1.1.3 MAGNESIUM SULFATE ($MgSO_4$)

The main reaction process when a magnesium sulfate solution attacks a Portland cement-based system is the formation of brucite $(Mg(OH)_2)$ and gypsum.

$$Mg^{2+} + SO_4^{2-} + Ca(OH)_2 + 2H_2O \rightarrow Mg(OH)_2 + CaSO_4 \cdot 2H_2O \quad (5.7)$$

As Portlandite is consumed in this reaction, the C-S-H phase will gradually decompose, forming an amorphous hydrous silica $(SiO_2 \cdot aq)$ and/or a magnesium silicate hydrate phase which is poorly crystalline $(3MgO \cdot 2SiO_2 \cdot 2H_2O)$.

The decomposition of C-S-H proceeding simultaneously with the formation of brucite and gypsum leads to a quite fast degradation of the concrete, which, in turn, leads to material softening and considerable strength reduction. This process also occurs in Portland cement-based systems with a low C_3A content, making magnesium sulfate more aggressive than sodium sulfate.

When C_3A is present, some ettringite formation can occur when exposed to magnesium sulfate solutions. However, this ettringite formation will only occur in more inward zones where the pH remains high enough. Since the amount of ettringite in this case is rather low, the concrete will be disintegrated due to C-S-H decomposition before significant swelling pressure can be formed.

Figure 5.29 provides an overview of the chemical reaction processes in the case of exposure to magnesium sulfate solutions, as summarized by Skalny et al. (2002).

5.3.2.1.1.4 THAUMASITE FORM OF SULFATE ATTACK (TSA)

In due presence of carbonate ions, thaumasite $(3CaO \cdot SiO_2 \cdot CO_3 \cdot SO_3 \cdot 15H_2O)$ can be formed directly from C-S-H in Portland cement-based systems, provided that the temperature is low enough (below 15°C) and that the pH levels are high enough (above 10.5) (Bassuoni and Nehdi 2009).

$$3Ca^{2+} + SiO_3^{2-} + CO_3^{2-} + SO_4^{2-} + 15H_2O \rightarrow$$
$$3CaO \cdot SiO_2 \cdot CO_2 \cdot SO_3 \cdot 15H_2O \quad (5.8)$$

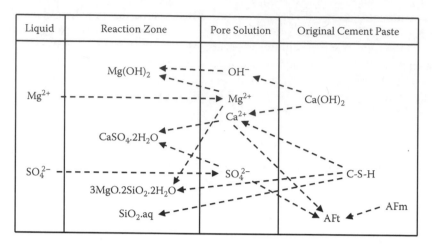

Figure 5.29 Overview of reactions for a Portland cement-based system in contact with a magnesium sulfate solution (after Skalny et al. 2002).

Thaumasite and ettringite have similar structures replacing $Al(OH)_6^{3-}$ ions with $Si(OH)_6^{2-}$ ions and $(3SO_4^{2-} + 2H_2O)$ with $(2CO_3^{2-} + 2SO_4^{2-})$ (Skalny et al. 2002). Nevertheless, in the case of direct thaumasite formation, the concrete loses strength due to decomposition of C-S-H and the consequent softening of the matrix rather than due to expansive cracking. TSA can proceed much faster when exposed to a magnesium sulfate solution, as in this case the C-S-H is decomposing faster.

Apart from the direct route to form thaumasite as explained in the previous paragraph, an indirect route through ettringite, called the woodfordite route, is also possible (Irassar 2009). However, according to Kohler et al., the woodfordite route is not followed, while the direct route is extremely slow, if not unlikely (Kohler et al. 2006). According to Kohler et al., thaumasite is rather formed through nucleation on the ettringite surface when the C-S-H is decomposed.

The appearance of thaumasite sulfate attack in real structures seems to be a point of discussion. However, in the 1990s, an increasing number of TSA cases were reported in the United Kingdom (Skalny et al. 2002). In a recent study, Bellmann et al. (2012) reported TSA to be the main damage cause in a series of 20 investigated structures, while ettringite only damaged one structure in this series.

The risk of TSA is said to increase when self-compacting concrete contains limestone filler, although this statement is also under debate. Bassuoni et al. (2009) reported some inferior performance of limestone filler-based self-compacting concrete ascribed to TSA upon exposure to sodium sulfate and cold temperatures. They stated that 'such a performance risk should be accounted for when qualifying SCC mixtures for use in environments conductive to TSA'. On the other hand, Irassar (2009) concluded that TSA on SCC is mainly governed by transport properties, and by previous damage caused by expansive ettringite formation. As the surface area accessible to sulfate ions increases substantially after expansive cracking due to ettringite formation, increasing dissolution of $Ca(OH)_2$ promotes instability of ettringite and decomposition of C-S-H, favouring thaumasite formation in the presence of carbonate and calcium ions. This latter view is supported by Schmidt et al. (2009), who conclude from their investigation that 'thaumasite formation was always preceded by expansion and cracking of the samples due to ettringite formation and given the very slow kinetics of thaumasite formation it was probably facilitated by the opening up of the structure due to ettringite induced cracking'. This might lead to the conclusion that TSA can occur in real structures, but that most probably it is only to be considered as a secondary damage mechanism, after previous damage by a more classical form of sulfate attack or any other damage mechanism leading to cracking.

5.3.2.1.1.5 SULFATE-ACID ATTACK

In some cases, e.g. exposure to sulfuric acid (H_2SO_4) or ammonium sulfate (($NH_4)_2SO_4$), a combination of sulfate attack and acid attack is obtained. Typical reaction products as in the case of sulfate attack will be formed, e.g. gypsum, in combination with acid attack. For this latter degradation mechanism, reference is made to Sections 5.3.3 and 5.3.4.

5.3.2.1.2 Influencing parameters

From the mechanism described in the previous section, important influencing parameters can be summarized as follows.

External chemical sulfate attack is promoted when sulfate ions can enter the concrete more easily. The *transport properties of concrete*, as also discussed in Chapter 1, thus significantly influence the process of sulfate attack. In terms of concrete mix design, this is significantly linked to the water/cement ratio.

Considering the chemical processes, the *cement chemistry* also plays an important role, especially the aluminate phase. Some sulfate reaction schemes are strongly linked to the presence of monosulfate and its conversion to ettringite.

A main influencing factor, of course, is the environment to which the concrete is exposed. Of major importance are the sulfate concentration and the type of sulfate solution. Furthermore, temperature and humidity also influence the sulfate attack processes.

5.3.2.1.3 Mitigation

A major measure to reduce the risk of sulfate attack is to produce a *dense concrete* with reduced transport properties. This can be achieved by providing a *low water/cement ratio*, as confirmed explicitly by Marchand et al. (2002) while theoretically analysing the effect of weak sodium sulfate solutions on the durability of concrete. In this way, sulfate ions will penetrate the concrete more slowly. However, care should be taken to avoid early-age (micro) cracking (e.g. due to autogenous shrinkage), which will, of course, increase transport properties in spite of the low water/cement ratio.

For sulfate attack cases in which expansive ettringite formation is the driving force leading to concrete damage, the application of a cement with *low C_3A content* is advised, or a so-called *high sulfate resisting (HSR) cement*. As no significant amount of monosulfate will be formed during hydration of an HSR cement, no subsequent transformation to ettringite can occur. In some standards, the restriction not only concerns C_3A, but also C_4AF, e.g. in ASTM C 150-2002. However, the application of an HSR cement does not help where the C-S-H is decomposed due to the action of the sulfate solution (e.g. in the case of magnesium sulfate).

Puzzolanic materials (fly ash, slag, silica fume, natural puzzolans) can also be helpful in reducing the risk of sulfate attack, although different materials might not be equally effective (depending on the alumina content, a low alumina content being beneficial) (Neville 2004). Their positive action goes along with the densifying effect on the pore structure reducing transport properties, with a reduction of the calcium hydroxide content of the hydrated system reducing gypsum formation, and with a reduction of the C_3A content reducing the risk of ettringite formation. However, in the case of puzzolanic materials, degradation by magnesium sulfate is more pronounced than degradation by other sulfate solutions. In the case of the addition of puzzolanic materials, an adequate curing of the concrete is of main importance because lack of curing might lead to a higher risk of (physical) sulfate attack in these cases (see further in Section 5.3.2.2).

Besides HSR cement, other *special cements* can significantly improve the resistance against sulfate attack such as calcium aluminate cement, phosphate cement, alkali silicate cement, and geopolymer cement (Skalny et al. 2002). A discussion of these special cement types, however, is beyond the scope of this textbook.

5.3.2.1.4 Example

In real structures, when expansive ettringite formation is the main cause of damage, a chemical sulfate attack can lead to map cracking as in the case for ASR. A major difference, however, is that cracks will typically not run through the aggregates, as opposed to ASR-induced cracks, which are initiated within the aggregates. While ASR can show a yellowish or brownish

Figure 5.30 Detail of a concrete balcony damaged by expansive chemical sulfate attack (courtesy of SCICON Worldwide bvba).

gel formation, this is also different for a sulfate attack, where a white discolouration can become visible due to the formation of gypsum. When sulfate attack leads to the decomposition of C-S-H, the concrete becomes soft and surface layers can be eroded.

An illustration of a concrete element internally damaged by chemical sulfate attack through expansive reactions is shown in Figure 5.30. The picture shows a concrete balcony while being rehabilitated, which contains many internal cracks. In the cracks, and in the interface zone around the aggregates, many crystals can be noticed. The expansive damage process was caused by sulfates released from (old type) bituminous materials which had been applied to make the balcony watertight.

5.3.2.2 Physical sulfate attack

5.3.2.2.1 Mechanism

The term 'physical sulfate attack' typically refers to the situation where partially buried concrete elements in sulfate-containing underground, or partially immersed in sulfate-containing water. It has been mentioned already that a strict distinction between 'physical' sulfate attack and 'chemical' sulfate attack is not realistic because many chemical processes typically occur together with salt crystallisation effects that give rise to the name 'physical attack'. More evidence of the chemical aspects will be given, but for reasons of clarity, the following paragraphs first explain the physical aspects.

When a concrete element is partially immersed in a sulfate solution, wick action will cause a transport of water through the porous concrete toward the drying face which is in contact with the surrounding air, which typically has a relative humidity below 100%. Wick action involves capillary suction, diffusion, and evaporation. Together with the water, sulfate (and other) ions are also transported into the concrete. On the drying surface, water evaporates and a salt concentration builds up near the surface because salts cannot be carried by water vapour. The high salt concentration near the surface causes back diffusion, leading to a nearly saturated pore solution zone near the surface (Pel et al. 2004).

In the case of a concrete element partially immersed in a sodium sulfate solution, due to wick action and back diffusion, a very high concentration of SO_4^{2-} and Na^{2+} ions is obtained in the pores near the drying surface, even higher than in the sodium sulfate solution itself (Liu 2010, Liu et al. 2012). Thanks to this super-saturation, crystallisation pressures could occur in the pores near the surface in the case of a reversible change of anhydrous sodium sulfate (thenardite) into decahydrate (mirabilite) (Neville 2004). This crystallisation process is considered to be a physical process, hence the name 'physical' sulfate attack. The part of the concrete element above the solution level is degraded significantly, showing efflorescence and cracking, while the immersed part typically remains nearly intact.

Quite often, a parallel is seen between physical salt (sulfate) attack in concrete and physical salt attack in bricks and natural stones. However, clear contradictions can easily be noticed (Liu 2010). Physical attack on concrete seems to increase with increased relative humidity of the surrounding air, while the opposite is noted in the case of bricks and stones. Furthermore, according to the basic principles of salt crystallisation in porous materials, a concrete with a higher water/cement ratio would have to show a better resistance against physical sulfate attack, while experiments and field cases show the opposite (Hime 2003).

These discrepancies illustrate that 'physical sulfate attack' of concrete can be heavily debated. According to Mehta (2000) efflorescence should not cause any damage, except under certain circumstances. This discussion has motivated Liu (2010) to study chemical aspects in 'physical sulfate attack'. He conjectures that the major cause for the distress of concrete partially exposed to a sulfate environment is more likely to be chemical sulfate attack, not salt weathering, salt crystallisation, or physical attack (Liu et al. 2011). This seems to hold both for sodium and magnesium sulfate solutions. Salt crystallisation, however, can play an additional aggravating role when exposed to sodium sulfate after damage initiation by chemical sulfate attack. The conclusion of the major role of chemical attack in the case of partially exposed structures seems to be in line with the findings of Bellmann et al. (2012), which are based on the field performance of 20 concrete structures.

Nevertheless, in the case of previous carbonation, crystallisation of sodium sulfate can occur in the cement paste and cause damage. Although

Liu did not find traces of salt crystallisation in concrete partially exposed to magnesium sulfate solution (Liu et al. 2010), epsomite ($MgSO_4 \cdot 7H_2O$) was reported by Gruyaert et al. (2012) in the case of partially exposed concrete which showed carbonation. Carbonation, thus, seems to enable and accelerate damage due to salt crystallisation in partially exposed concrete elements. This illustrates the importance of adequate curing and reducing carbonation for the sulfate resistance of concrete structures, especially in the case of the addition of puzzolanic materials.

It could be concluded that 'physical' sulfate attack, causing damage due to salt crystallisation, is a real damage mechanism. However, it only plays an aggravating role after the concrete is first damaged by chemical sulfate attack or in case of previous carbonation. For many field cases ascribed to 'physical' sulfate attack, 'chemical' sulfate attack seems to be the main damage mechanism.

Nevertheless, there might be some specific circumstances where the action is purely physical and could be produced by salts other than sulfate, justifying the name 'physical attack' (Neville 2004), but sometimes also called 'salt weathering' or 'salt hydration stress'. The resulting damage has a similar appearance to surface scaling, as in the case of freeze–thaw cycles (Haynes and Bassuoni 2011). This process could be favoured by cooling and heating and wetting and drying, with cyclic crystallisation of salts (Stark 2002).

5.3.2.2.2 Influencing parameters

As the main damage mechanism is the chemical sulfate attack, the influencing parameters are the same as mentioned in Section 5.3.2.1.2. In the case of pure crystallisation effects, wetting and drying and cooling and heating cycles will accelerate the damage process.

5.3.2.2.3 Mitigation

The same mitigation measures can be mentioned as in Section 5.3.2.1.3. Additionally, a *coating* could be applied to the concrete drying surface exposed to the surrounding air in order to stop or limit wick action. Another option is to *isolate the concrete from the soil* (Haynes and Bassuoni 2011).

5.3.2.2.4 Example

As an important part of the 'West Development Strategy', China will make huge investments in railway construction and will more than double the existing operating mileage of railways in its western regions, bringing it to fifty thousand kilometres by 2020 (Liu 2010). However, in the western regions of China, the groundwater shows a high content of sulfates

Figure 5.31 'Physical' sulfate attack in a concrete tunnel in the Chengkun Railway link, China (courtesy Z. Liu).

including Na_2SO_4, $CaSO_4$, $FeSO_4$ and $MgSO_4$. For example, in Korla City, Xinjiang Province, China, within the groundwater the SO_4^{2-} content is about 21,000 mg/l, while the Mg^{2+} content is about 3600 mg/l. In the Baijialing Tunnel of Chengkun Railway (from Chengdu City to Kunming City), the SO_4^{2-} concentration in the water present in the tunnel drainage channel can reach as high as 32,475 mg/l. In this severe, sulfate-rich environment, physical sulfate attack in railway tunnels is a major durability issue, as is shown in Figure 5.31, which shows an existing tunnel of the Chengkun Railway link.

5.3.2.3 Delayed ettringite formation (DEF)

Delayed ettringite formation (DEF) is an internal form of sulfate attack which can occur at later age in the case of heat cured concrete elements. A confusing terminology exists in literature and a better name would be 'heat-induced sulfate attack', as suggested by Skalny et al. (2002). Nevertheless, the phenomenon is much better known as DEF.

When Portland cement-based concrete is cured at normal temperatures, calcium sulfate reacts with calcium aluminate and water, producing ettringite, which during the hydration process is transformed to monosulfate. In the case of heat curing at temperatures above 60 to 70°C, this normal reaction scheme is not followed in the same way. Ettringite will not be formed or will not be stable at high temperature, and sulfate will remain available in the system, adsorbed by C-S-H, together with monosulfate. At a later age, after cooling down, 'delayed' ettringite formation can occur by reaction of monosulfate and the internally available sulfate which is released from the C-S-H. This delayed ettringite formation can be expansive and cause cracking.

Many different theories exist to fully explain the detailed mechanism of DEF, as discussed by Skalny et al. (2002). It is also discussed whether DEF is really causing damage to the concrete, as not all ettringite formation leads to the expansion of the cement paste. According to many researchers, pre-existing cracks are needed before delayed ettringite formation can occur, e.g. caused by freeze–thaw, ASR, or thermal cracking. In this view, the expansion caused by DEF is not significant and there would be no damage if no other mechanism had caused pre-cracking.

While scientific debate is still going on, appropriate measures to avoid damage due to DEF can be taken and are sometimes provided in standards. DEF typically concerns the concrete precast industry where, for reasons of production speed, concrete elements quite often are steam cured or heat cured in another way. During hydration, temperatures higher than 60° to 70°C should be avoided.

Also in hot climates, such as those in the Middle East, DEF can be a matter of concern because the hydration process typically occurs at higher temperatures, especially in the case of massive elements in which the heat of hydration adds to the environmental temperature effect. The risk of DEF increases with increased curing temperature.

Although the mechanism of DEF is still under debate, it seems that it does not cause damage in the case of blended cements such as blast furnace slag cement. Nevertheless, a clear correlation between the occurrence of DEF and the type and composition of the cement is not currently available.

5.3.2.4 Sea water

When discussing sulfate attack, sea water is often referred to as a case of concern in this respect. It is true that sea water contains sulfate ions, SO_4^{2-}, but many other ions are also present, including Na^+, Mg^{2+} and Cl^- to only name the main ones.

Standard provisions typically require a high sulfate resisting (HSR) cement for concrete structures in contact with sea water. However, this is probably not the best choice to reduce the risk of chloride-induced reinforcement corrosion, because the chloride binding capacity in the case of HSR cement is reduced, having a lower content of calcium aluminate. As mentioned by Skalny et al. (2002), 'the formation of ettringite in sea water attack typically does not lead to expansion and cracking of the concrete and it is believed that the formation of this phase is non-expanding in the presence of excessive amounts of chloride ions'. It can be noted that in many countries chloride-induced reinforcement corrosion (see Section 5.4.3) is causing more damage to marine structures than sulfate attack.

It is to be remarked, however, that sea water can be very different all over the world. In some locations, very high concentrations of some specific ions can cause severe damage to exposed concrete structures, sometimes accelerated by a higher temperature in hot climates. Due attention should be paid to the local exposure conditions when dealing with marine structures.

5.3.3 Acid attack

As concrete is a calcium-rich alkaline environment, it will react with nearly all possible acids. The solubility of the reaction products will be of major importance for the resulting degradation level of the concrete, especially the solubility of the calcium salt, which is typically the first salt to be formed in case of an acid attack. This also explains why the rate of acid attack quite often more significantly depends on the level of water movement near the concrete surface, rather than on the type of cement or aggregate. Nevertheless, the latter parameters surely have an influence (De Belie 1997). It is often considered that cement types with a lower calcium content will be more resistant to acid attack. Although this often seems the case, it cannot be fully generalized to all cases. In some cases, a low $Ca(OH)_2$ content in the matrix can lead to a lower buffer capacity and a faster degradation of the C-S-H phase. A similar reasoning can be followed concerning the aggregates.

The main effect of the acid is the dissolution of the cement matrix, leading to a degradation of the concrete. In contrast with typical cases of sulfate attack, the degradation process in the case of acid attack is not linked to any significant expansion reaction because ettringite and thaumasite are not stable in the acid environment. The level of degradation largely depends on the type of acid to which the concrete is exposed. Furthermore, it is important to mention that in real exposure cases wastewaters can contain a multitude of different acids, which makes the degradation process very complex. The resulting degradation cannot be predicted by simply considering a superposition of the degradation caused by each acid separately.

Larreur-Cayol et al. (2011) studied the effects of three different organic acids typically present in agro-industrial wastewaters, namely oxalic, citric, and tartaric acids, and made a comparison with the effect of acetic acid. The specimens immersed in citric acid showed large amounts of white salts on their surfaces, which, however, could be removed very easily. A significant and rapid erosion of the outer part was also noticed. In the case of acetic acid, there was no clear salt precipitation, which is in accordance with the good solubility of calcium acetate. Some cracking was observed, probably caused by shrinkage due to the loss of calcium. Samples immersed in tartaric acid did not show visible degradation during the first month; but, after one month, a layer of very loose yellowish salt precipitated on the surface. After two months of exposure, the outer layer of the samples gradually dissolved. Very little attack was noticed on the specimens immersed in oxalic acid. This comparative study clearly showed that the most important parameters leading toward aggressiveness of acids is the solubility of their salts. Furthermore, the molar volume of the salts is also a key parameter, showing a better protective effect in the case of higher molar volume. Due to the combination of low solubility of the salt and suitable molar volume, the salt formation can seal the capillary pores, preventing or at least slowing further attack. The way the salt precipitates on the matrix, and the salt's affinity for the matrix are significant influential parameters (Larreur-Cayol et al. 2011).

As for the mechanism of acid attack, most typically a sequential process is going on, as illustrated by the existence of different layers after some period of exposure. In the bulk of the concrete, a sound zone is noticed, with no signs of attack. More toward the surface, a first layer surrounding the sound bulk material typically shows signs of decalcification. Depending on the type of acid, other chemical changes can occur. In a second layer, which is a thin outer surface layer, a more severe degradation can be noticed, involving advanced dissolution of the C-S-H phase. The severely degraded outer layer could easily be eroded by flowing water or by other abrasive actions.

Boel et al. (2008) applied X-ray computed microtomography to study the degradation process of a small cement paste sample (diameter 5.5 mm) exposed to lactic and acetic acid. The paste, representative for limestone filler-based self-compacting concrete, was composed of 300 kg/m³ blast furnace slag cement CEM III 42.5 LA, 300 kg/m³ limestone powder, 165 kg/m³ water and 1.8 l/m³ super plasticizer. The sample was immersed in a solution of lactic and acetic acids, leading to a disintegration of the cement paste. By means of X-ray microtomography, the sample was scanned after 25 hours and after 21 days of exposure. The resulting images are shown in Figure 5.32. After 25 hours of exposure, a low density outer zone was

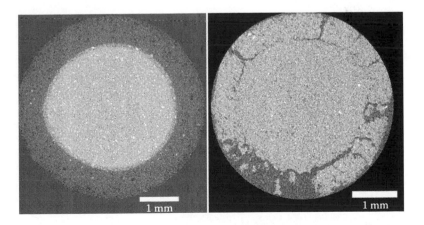

Figure 5.32 X-ray computed microtomography scan of paste sample exposed to lactic and acetic acid, after 25 h (left) and 21 days (right) of exposure (Boel et al. 2008).

formed around the undamaged core of the sample. After 21 days, the outer part showed cracks and was partially released. The front of the attack is clearly visible in the sample after 21 days of exposure.

One typical case of severe attack is caused by sulfuric acid, H_2SO_4, which, in fact, is a combination of acid attack and sulfate attack (Skalny et al. 2002). In the case of exposure to sulfuric acid, calcium hydroxide will react with the acid, forming gypsum (calcium sulfate). A similar reaction occurs, degrading the C-S-H phase. Furthermore, because of the reduced pH value due to the presence of the acid, monosulfate and ettringite will become unstable and form gypsum and aluminium sulfate. In the case of running fluid near the exposed concrete surface, the degrading concrete surface can be eroded very easily.

A layered attack process in the case of sulfuric acid attack was clearly noted by Boel et al. (2008) by means of X-ray computed microtomography on a paste sample (with composition as mentioned before) immersed in sulfuric acid. Figure 5.33 shows the exposed sample after 22 hours and 21 days of exposure. After 22 hours, only an outer layer (thickness 0.15 mm) was visibly attacked and pushed off, while the rest of the sample showed no noticeable damage. After 21 days, the damaged outer zone reached a thickness of 0.45 mm. However, the damaged zone seemed to consist of two separate layers with an air layer in between.

Figure 5.34 shows a well advanced case of sulfuric acid attack on a concrete wastewater tank in an industrial plant. The concrete cover disappeared over a large area, making the corroded reinforcement fully visible. In this case, river gravel was used as an aggregate, which resists the acid environment quite well. As it is the cement matrix which was attacked, the damaged concrete has a 'washed-out' appearance.

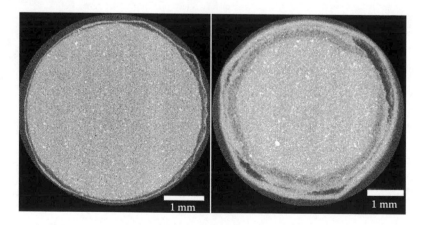

Figure 5.33 X-ray computed microtomography scan of paste sample exposed to sulfuric acid, after 22 h (left) and 21 days (right) of exposure (Boel et al. 2008).

Figure 5.34 Sulfuric acid attack on a concrete wastewater tank in an industrial plant (courtesy of SCICON Worldwide bvba).

A special case of sulfuric acid attack can be found in sewer systems, involving the action of micro-organisms or bacteria. This is the so-called biogenic sulfuric acid attack, which will be explained in Section 5.3.4.

Ammonium salts are also reported to be very aggressive. For ammonium chloride and ammonium sulfate, the occurring chemical reactions are as follows.

Ammonium chloride:

$$2\ NH_4Cl + Ca(OH)_2 \rightarrow CaCl_2 + 2\ H_2O + 2\ NH_3\uparrow \qquad (5.9)$$

$$2\ NH_4Cl + CaCO_3 \rightarrow CaCl_2 + H_2O + CO_2 + 2\ NH_3\uparrow \qquad (5.10)$$

Ammonium sulfate:

$$(NH_4)_2SO_4 + Ca(OH)_2 \rightarrow CaSO_4 \cdot 2H_2O + 2\ NH_3\uparrow \qquad (5.11)$$

$$x\ (NH_4)_2SO_4 + x\ H_2O + x\ CaO \cdot SiO_2 \cdot aq \rightarrow$$
$$x\ CaSO_4 \cdot 2H_2O + SiO_2 \cdot aq + 2x\ NH_3\uparrow \qquad (5.12)$$

Upon the reaction with $Ca(OH)_2$, gaseous ammonia, NH_3, is formed, which will escape from the concrete. The pH value of the system is reduced and monosulfate as well as ettringite become unstable. In the case of ammonium sulfate, a reaction with the C-S-H phase also occurs, converting it to amorphous hydrous silica (Skalny et al. 2002). Due to this reaction, a strength loss is obtained.

When carbon dioxide dissolves in water, carbonic acid, H_2CO_3, is formed. Upon reaction with a cementitious system, calcium bicarbonate, $Ca(HCO_3)_2$, is formed, which can go into solution. In some cases that involve contact with flowing water containing high amounts of carbon dioxide, the cement stone can be eroded.

In general, acid attacks on concrete elements are nearly impossible to avoid. However, some measures can be taken to slow the degradation process. The most commonly mentioned measure to reduce or delay degradation resulting from acid attack is to provide a dense concrete matrix by lowering the water/cement ratio. Another option is to provide a concrete liner or coating to avoid contact between the acid and the cementitious material.

5.3.4 Biogenic sulfuric acid attack

In concrete sewer systems, it is quite often noticed in practice that severe damage occurs after a very short time, sometimes only a few years. Figure 5.35 shows the remarkable case of a concrete sewer which after

Figure 5.35 Biogenic sulfuric acid attack in a concrete sewer system, totally damaging the upper part of the concrete pipe above the wastewater line.

less than 5 years had almost entirely disintegrated. At the very least, it
can be observed that the part of the concrete pipe above the wastewater
line is heavily damaged and even disappeared over significant lengths. This
case was caused by so-called biogenic sulfuric acid attack, which is more
aggressive than pure sulfuric acid attack (see Section 5.3.3). The adjective
'biogenic' is added to make clear that micro-organisms or bacteria play an
important role in the damage process.

Figure 5.36 schematically shows the different steps involved in the
damage process. In wastewater, sulfate ions (SO_4^{2-}) are typically present. By the action of anaerobic bacteria (bacteria which can survive in
environment without oxygen, as is typically the case in wastewater), the
sulfate ions are converted into sulfide ions (S^{2-}), which further react with
hydrogen (H^+) to form sulfide gas (H_2S). This sulfide gas escapes from the
wastewater, entering an oxygen containing atmosphere above the water
line in the sewer pipe. On the concrete surface above the water line within
a thin moisture layer, the hydrogen sulfide gas is dissolved. In this environment on the concrete surface aerobic *Thiobacillus* bacteria convert the
hydrogen sulfide to sulfuric acid (H_2SO_4). Although it is clear that the
sulfuric acid is damaging the concrete, some debate exists on the exact

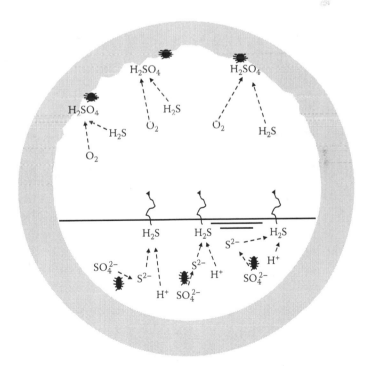

Figure 5.36 Schematic representation of the process involved in biogenic sulfuric acid
attack.

mechanisms. It is not fully clear whether the *Thiobacillus* bacteria migrate into the concrete, producing sulfuric acid much closer to the damage front, or whether the produced sulfuric acid itself is moving inward (O'Connell et al. 2010). Based on a comparison with pure sulfuric acid attack where the penetration of sulfuric acid is somewhat obstructed due to the damage process, an active contribution by the *Thiobacillus* bacteria seems plausible in explaining the much faster damage process in the case of biogenic sulfuric acid attack.

It is clear that the performance of concrete with regard to biogenic sulfuric acid attack cannot be fully judged by purely chemical tests alone because the microbial effects are not considered in this way. Tests in a real biological environment will enable more reliable conclusions concerning the concrete performance in sewer pipes (Monteny et al. 2000).

According to the review performed by O'Connell et al. (2010), there seems no agreement on the parameters of significance governing the resistance of concrete to biogenic sulfuric acid attack. A dense concrete is typically considered to perform better because of reduced penetration properties, although a refined pore structure could also increase capillary suction. Debate also exists concerning cement and aggregate types and contradictory results can be found in literature. Although limestone aggregates can be attacked by the acid, it seems that due to their buffer capacity they can delay the concrete degradation much better than river gravel.

According to Leemann et al. (2010), who studied concrete degradation due to the effect of bacteria oxidizing ammonium from nitrate, it is also important to notice that calcite precipitation close to the surface, leading to the formation of some dense layer, plays an important role in addition to the dissolution of hydrates. They experimentally found that concrete deterioration in the case of biogenic acid attack by the nitrifying biofilm is correlated with the CaO content of the cement. In the case of a higher CaO content, more calcite is formed offering a better protection against acid attack. It is not clear to what extent this conclusion also holds for biogenic sulfuric acid attack.

5.4 REINFORCEMENT CORROSION

5.4.1 General

Corrosion of reinforcing steel probably is the most widely spread damage mechanism in concrete structures. Typical features of corroding concrete structures are the occurrence of cracks along the reinforcing steel and spalling of the concrete cover, which leave the corroding reinforcing steel visible. Brown colouring due to outflowing corrosion products can often be noticed as well.

It is well known that when steel is exposed to normal environmental conditions (at least in most places on earth), it will corrode because of the reaction of iron (which is the main component of steel) with water and oxygen. Corrosion of steel is a complex electrochemical reaction process with many influencing factors. For a detailed treatment of general corrosion processes, reference is made to specialised literature. More simply, it can be said that the corrosion process requires an anode where electrochemical oxidation takes place and a cathode where electrochemical reduction takes place. Furthermore, negative and positive particles have to be transported through the electrically conducting corroding steel and the surrounding electrolyte solution.

When steel is placed in a water solution (electrolyte), a certain number of iron atoms will go into solution as iron ions, leaving some electrons behind in the steel:

$$Fe \rightarrow Fe^{2+} + 2e^- \tag{5.13}$$

This is the anode reaction or oxidation, occurring at the anode (negative pole). The electrons (e⁻) will be transported through the steel toward a cathode (positive pole), where a cathode reaction or reduction takes place, which in the presence of water and oxygen can be described as follows:

$$O_2 + 2H_2O + 4e^- \rightarrow 4OH^- \tag{5.14}$$

A classical illustration of a steady corrosion process is the electrochemical macrocell, consisting of two different metals which are electrically connected and immersed in an electrolyte solution, as illustrated in Figure 5.37. One of the metals becomes the anode, while the other becomes the cathode. At the anode, metal ions go into solution. Thus, this metal is oxidizing or corroding. The other metal remains sound. Here, the electrons coming from the corroding metal react with oxygen and water, forming hydroxyl ions OH⁻. The macrocell clearly illustrates that a corrosion process is the combination of chemical transformations and the transfer of electrical charges, justifying the term 'electrochemical' process.

In the case of steel reinforcement in concrete where the concrete pore solution acts as an electrolyte, normally no second type of metal is present except in case of cathodic protection, which however is beyond the scope of this textbook. Nevertheless, anodes and cathodes can occur on the same steel surface, due to heterogeneities in the steel (metallurgical segregation or different grain orientations), or due to local variations in the electrolyte (variations in concentration or aeration) (Rosenberg et al. 1989). As anode and cathode can be close together, the term microcell is typically used instead of macrocell.

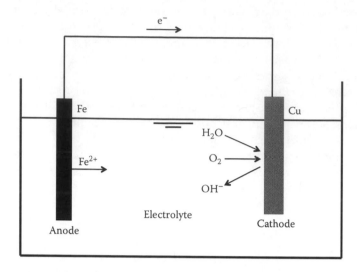

Figure 5.37 Macrocell representation of a corrosion process.

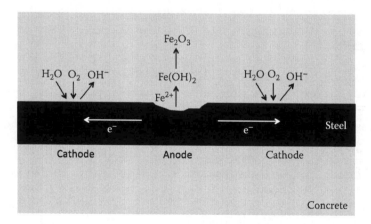

Figure 5.38 Corrosion microcell of steel in a cementitious system.

Figure 5.38 shows a typical microcell for corroding steel in a cementitious system. At the anode, iron ions go into solution. The remaining electrons are transported toward the cathode, where in the presence of water and oxygen some hydroxyl ions are formed. These hydroxyl ions are transported through the electrolyte solution and react with the iron ions forming iron hydroxide $Fe(OH)_2$, which can be further oxidized to iron oxide Fe_2O_3. The expansive formation of corrosion products leads to tensile stresses in the surrounding concrete, finally resulting in cracking and spalling of the concrete cover.

A general distribution of anodes and cathodes on the steel surface combined with a change of their position during the corrosion process, will lead to a quite uniform corrosion attack on the steel reinforcement. However, if a small number of anodes is located at fixed points, localised corrosion can occur (Rosenberg et al. 1989).

Considering the given information, the layman may question why steel reinforcement is added to concrete containing oxygen and water, which potentially leads to corrosion. The reason reinforcing steel in concrete normally does not corrode is because the pore solution in the concrete is totally different from a typical aqueous environment (water, rain) on the earth's surface. The pore solution in a cementitious material typically has a high pH value because of the high content of calcium hydroxide resulting from cement hydration (see Chapter 1). The influence of the pore solution, or any electrolyte in general, on the steel corrosion process is illustrated by a so-called Pourbaix diagram, as shown in Figure 5.39.

Based on thermodynamic information, a Pourbaix diagram shows the conditions of electrochemical potential and pH value of an electrolyte where corrosion is expected or where immunity or passivation is reached (Pourbaix 1976). The electrochemical potential (expressed in Volts, V) is the potential difference between the anode and a reference electrode, and is arbitrarily defined as zero for the hydrogen electrode which is used as a reference for all metal potentials (Rosenberg 1989). In the given Pourbaix diagram, which is a simplified diagram for iron in an aqueous environment, a typical natural environmental condition illustrates that iron (steel) can

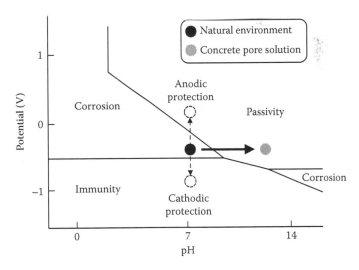

Figure 5.39 (Simplified) Pourbaix equilibrium diagram for iron in aqueous environment.

actively corrode in the presence of water and oxygen. However, in the case of steel embedded in concrete, the higher pH value of the electrolyte brings the steel into a passivation area. The reinforcing steel will be passivated and will not actively corrode.

The passivation of steel in a concrete pore solution can be explained by the presence of a very high content of hydroxyl ions, resulting from the hydration process. Iron ions going into solution react with the hydroxyl ions forming iron hydroxide. In the presence of a high content of hydroxyl ions, iron hydroxide precipitates on the steel surface, providing a thin protective film that obstructs further dissolution of iron ions. The steel is now passivated, due to the presence of a passivating layer of iron hydroxide. In the case of a lower pH value of the electrolyte, the precipitated layer of iron hydroxide becomes more porous and more permeable, so that the corrosion process can go on actively. Consequently, it can be said that the protection against corrosion of steel reinforcement in concrete is provided by the high alkalinity of the pore solution. Other measures can be provided to protect against steel corrosion, as will be briefly explained further on.

Knowing the influence of the alkalinity of the pore solution, the normal situation of steel reinforcement in concrete is a condition of passivation. Reinforcing steel in concrete is not expected to corrode actively. However, seeing the widely spread corrosion damage to many reinforced concrete structures, it is clear that some mechanisms can counteract the protective role of the high alkalinity of the pore solution. The protective passivating layer on the steel surface can be attacked by carbon dioxide (CO_2) and/or chloride ions penetrating the concrete from the environment. This process is called depassivation. Corrosion due to penetration of carbon dioxide, e.g. present in the air, is called *carbonation induced corrosion,* while *chloride induced corrosion* refers to the effect of penetrating chloride ions, e.g. ions present in sea water or de-icing salts. Both mechanisms will be further detailed in the following paragraphs.

Considering the required depassivation of the reinforcing steel before active corrosion can occur and potentially cause damage to the concrete, a two-stage corrosion process of reinforcing steel in concrete is typically considered while studying the service life, as illustrated in Figure 5.40, often referred to as the Tuutti model (Tuutti 1982). During the *initiation phase,* the reinforcing steel will be gradually depassivated by the penetration of carbon dioxide and resulting carbonation of the cement matrix or by the penetration of chloride ions as already mentioned. As soon as the reinforcing steel is depassivated, active corrosion can proceed during the *propagation phase.* The corrosion rate depends on many influencing factors, of which the availability of water and oxygen are of main importance.

Depending on the defined corrosion acceptance level, the service life of the structure will be the sum of the initiation time and the propagation time. In the stage of durability design, quite often the end of the initiation

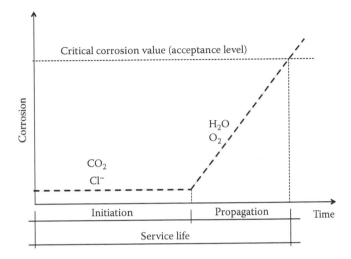

Figure 5.40 Phenomenological evolution of the corrosion process of reinforcing steel in concrete.

phase is defined as the end of the service life of the concrete structure, and no active corrosion at all is accepted. On the one hand, this approach is often justified because the propagation time is typically much shorter than the initiation time (in the case of appropriate design and execution of the structure). On the other hand, it seems somewhat too conservative in general to define the end of service life before any real damage occurs. However, it is also important to be aware that most corrosion problems in real concrete structures occur due to insufficient concrete quality and/or insufficient cover thickness, substantially reducing the initiation time (see also Sections 5.4.2 and 5.4.3). Thus, it seems justified to focus indeed on the initiation time, duly including quality assurance procedures to make sure that the as-built structure meets the designed properties (concrete quality, concrete cover). Keeping the reinforcing steel passivated during the entire service life is a good and safe design approach, but appropriate quality procedures are also needed during execution to make sure that design conditions are reached in practice.

Referring again to the Pourbaix diagram, it could be concluded that bringing the steel reinforcement into an environment with a higher alkalinity is an efficient measure to prevent corrosion. The fortunate situation in case of reinforced concrete is that this kind of alkaline environment is automatically provided by the hydration process of the cement. However, as already mentioned and as will be detailed further on, the protective passivating layer can be attacked due to external actions. Other, more active corrosion control measures can be applied based on the information given in the Pourbaix diagram. While not changing the pH value (around 7), protection can be obtained by bringing the potential into the immunity region (i.e. lowering the potential). This is called *cathodic protection.* Another option is to increase the potential to bring the steel into the passivity region, which is called *anodic protection.* In the case of concrete structures, cathodic protection is a well-known way to mitigate corrosion of the steel reinforcement. A detailed treatment of these protection techniques, however, is beyond the scope of this textbook.

Other general corrosion control measures include the application of *corrosion resistant alloys*; these typically contain chromium. On the surface of these special alloys, an effective and resistant passivating layer is developed. However, these alloys are typically expensive and are only applied in very special cases. Another option is to provide *coatings* on the reinforcing steel, which provide a barrier between the steel and the corrosive environment. *Corrosion inhibitors* can also be applied to the steel surface or mixed into the concrete. They can have different actions such as forming a protective layer on the steel surface or providing a buffer action in the surrounding electrolyte, e.g. capturing chloride ions.

5.4.2 Carbonation-induced corrosion

As illustrated in the Pourbaix diagram, the passivation layer of the reinforcing steel will become unstable in the case of the dealkalization of the surrounding cement paste. Typically, a threshold pH value of 8 to 9 is needed to reach overall depassivation of the reinforcement steel. This can be caused by the penetration of acids into the concrete, which will neutralize the alkaline environment. For typical concrete structures, the most important acid causing depassivation of the steel reinforcement is atmospheric carbon dioxide CO_2. It is commonly present in the air where it has an average concentration of 0.03 vol%, but can reach levels as high as 0.3 vol% in large cities. The carbon dioxide will penetrate the concrete and chemically react with alkaline elements (mainly $Ca(OH)_2$, but also $Na(OH)$ or $Ka(OH)$) to form carbonates. This process, commonly called carbonation, can be described by the following simplified chemical reactions:

$$Ca(OH)_2 + CO_2 \rightarrow CaCO_3 + H_2O \tag{5.15}$$

$$2NaOH + CO_2 \rightarrow Na_2CO_3 + H_2O \tag{5.16}$$

$$2KOH + CO_2 \rightarrow K_2CO_3 + H_2O \tag{5.17}$$

In somewhat more detail, it is to be noted that the carbon dioxide first dissolves in the pore solution according to the following reaction scheme (with g and aq standing for gaseous and aqueous respectively):

$$H_2O + CO_2 \ (g) \rightarrow HCO_3^- \ (aq) + H^+ \ (aq) \tag{5.18}$$

$$HCO_3^- \ (aq) \rightarrow CO_3^{2-} \ (aq) + H^+ \ (aq) \tag{5.19}$$

Afterwards, a neutralisation reaction will proceed, according to the following reaction, in the case of the carbonation of $Ca(OH)_2$:

$$Ca^{2+} \ (aq) + 2OH^- \ (aq) + 2H^+ \ (aq) + CO_3^{2-} \ (aq) \rightarrow CaCO_3 + 2H_2O \tag{5.20}$$

By this chemical reaction, hydroxyl ions will be removed from the pore solution, reducing the pH value. Furthermore, the formed calcium carbonate, $CaCO_3$, which has very low solubility, will precipitate in the pores. In Portland cement-based systems, the carbonation process will lead to a reduced porosity of the concrete and further reduce penetration of carbon dioxide. However, in blended systems, e.g. containing blast furnace slag, an increased porosity can be noticed due to carbonation.

While at first instance, the penetrating CO_2 reacts with the calcium hydroxide $Ca(OH)_2$, other hydration phases such as C-S-H can also be carbonated in the case of higher CO_2 concentrations. Figure 5.41 shows the evolving carbonation front in concrete, by considering the CO_2 concentration and the pH value as functions of the distance to the exposed concrete surface. As long as the reinforcing steel is within the non-carbonated zone,

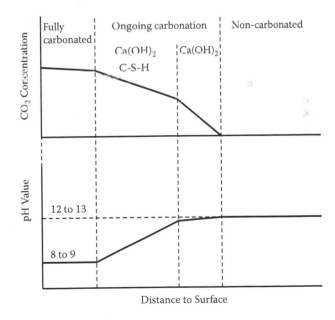

Figure 5.41 Evolving carbonation front in concrete: CO_2 concentration and pH value as a function of the distance to the exposed concrete surface.

it remains passivated. The thickness of the concrete cover thus needs to be appropriately determined, based on the penetration rate of CO_2, in order to protect the steel rebars from corrosion.

One of the main influencing factors in determining the penetration rate of carbon dioxide is the moisture condition of the concrete. Diffusion of CO_2 proceeds at a much higher rate (by a factor of about 10,000) in air than in water. Saturated concrete (e.g. part of quay walls permanently under water) will hardly carbonate because diffusion of carbon dioxide in saturated concrete is extremely slow. On the other hand, fully dry concrete will also not carbonate because water is needed for the carbonation reaction as explained before. Thus, carbonation of concrete depends on the possibility of penetration by carbon dioxide (which goes faster in dry conditions) and the ability of carbon dioxide to go into solution as a first step in the chemical carbonation process (which, of course, is not possible in fully dry conditions). Consequently, the highest carbonation rates are reached at intermediate humidity levels (50% to 80% relative humidity of the surrounding air), as illustrated in Figure 5.42 (Rosenberg et al. 1989).

The diffusion of carbon dioxide in steady-state conditions can be generally described by Fick's first law, expressing that the diffusion rate, J (in mol/m²s), is proportional to the concentration gradient dc/dx (c being the concentration, in mol/m³, and x the one-dimensional coordinate, in m) and to the diffusion coefficient D (in m²/s):

$$J = -D \ (dc/dx) \tag{5.21}$$

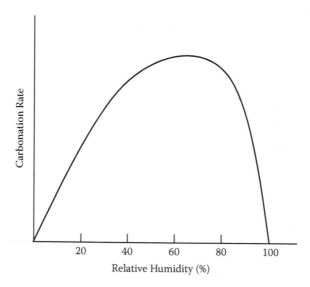

Figure 5.42 Influence of moisture conditions on the carbonation rate in concrete.

In real concrete, steady-state conditions are not reached, as the carbon dioxide concentration is a function of time. This has to be considered by following Fick's second law, giving the concentration as a function of time:

$$\frac{\partial c}{\partial t} = \frac{\partial}{\partial t}\left(D\frac{\partial c}{\partial x}\right)$$
(5.22)

The diffusion coefficient D can also be time and location dependent. However, in the case of a constant diffusion coefficient, considering an environmental carbon dioxide concentration equal to c_0, and an initial carbon dioxide concentration within the concrete equal to zero, the solution of Fick's second law is obtained by the following equation (with erf being the error function):

$$c = c_0\left(1 - erf\left(\frac{x}{2\sqrt{Dt}}\right)\right)$$
(5.23)

When transforming this equation in order to find the penetration depth as a function of the concentration, it is found that:

$$x = 2\sqrt{D}\ erf^{-1}\left(1 - \frac{c}{c_0}\right)\sqrt{t}$$
(5.24)

This equation shows that the penetration depth of carbon dioxide is a function of the square root of time. This was mentioned earlier in Chapter 1, Section 1.4.6, where the importance of the cover thickness for reaching the service life was stressed. Reducing the cover thickness by half reduces the corrosion initiation time by significantly more than half! It is to be noted, however, that the obtained square root time relation is based on a constant diffusion coefficient, valid for non-carbonated concrete. The reaction of the carbon dioxide with the hydration phases and the resulting influence on the pore structure, is not considered. Neither is the influence of the moisture content of the concrete, which will typically have a major impact on the diffusion coefficient as mentioned before. Further changes can be made to the given equation in order to take these effects into account (Audenaert 2006).

In real conditions, it seems that after time the depth of the carbonation front will be somewhat lower than predicted by the square root time relation, as illustrated in Figure 5.43. An ultimate value after a very long time is even suggested, depending on the density of the concrete, the amount of carbonatable material, and the humidity of the environment (Schiessl 1976).

The carbonation of concrete also depends on the type of binder. In the case of blast furnace slag cement, part of the $Ca(OH)_2$ produced by the

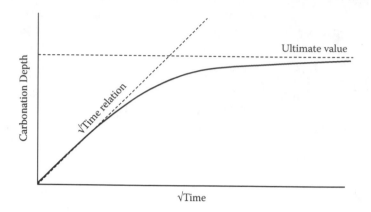

Figure 5.43 Evolution of carbonation depth in real conditions.

Portland hydration will be consumed by the reaction of the slag. This leads to a reduced $Ca(OH)_2$ content, and thus less carbonatable material. On the other hand, a more dense pore system can be obtained, although the reactions proceed more slowly. Nevertheless, the final pore structure depends heavily on the curing conditions. Insufficient curing will lead to more open pore structures in the case of binders containing blast furnace slag. The precise effect on the carbonation process is difficult to predict, and literature shows a large variation in results. In this context, it should be mentioned that curing conditions in laboratory research are typically much better than in real conditions. In general, it can be expected that in real concrete structures containing blast furnace slag, carbonation proceeds significantly more rapidly than in the case of pure Portland-based concrete. In the case of cement replacement by fly ash or other puzzolanic materials, similar phenomena can be found as for blast furnace slag, typically leading to increased carbonation rates and depths. However, in the case of the additional application of fly ash without cement reduction, a more dense structure can be obtained which results in a reduced carbonation rate.

Carbonation of concrete itself does not cause any real damage to the concrete structure. It only provides the conditions required for reinforcement corrosion to occur. As soon as the steel reinforcement is depassivated, corrosion can propagate. The corrosion rate depends on the availability of water and oxygen. The highest corrosion rates are obtained in the case of cyclic wetting and drying, e.g. due to tidal effect or rain. Permanently wet concrete, e.g. concrete under water, will not corrode because the diffusion of oxygen is extremely slow.

In the case of a constant relative humidity in the environment, the highest corrosion rates will be found for a relative humidity around 95% with decreasing corrosion rates for decreasing relative humidity, as illustrated in

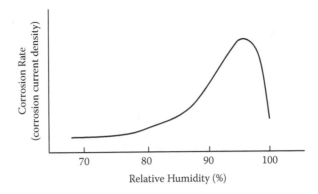

Figure 5.44 Corrosion rate in carbonated concrete as a function of the relative humidity of the environment.

Figure 5.44. This decreasing corrosion rate is also linked to the increasing electrical resistivity for decreasing moisture contents of the concrete. In a relatively dry environment (e.g. 50% to 60% relative humidity) where concrete carbonation can be relatively fast as explained before, the steel reinforcement will not show significant corrosion even if it lies in the carbonated zone. This illustrates that in some cases it is too conservative to only consider the initiation time and neglect the propagation time when studying the remaining service life of a concrete structure.

Typical examples of reinforced concrete structures affected by carbonation-induced corrosion can be found in residential and office buildings in many areas, as illustrated in the case of a concrete column of an office building shown in Figure 5.45. During the corrosion process, expansive formation of corrosion products leads to internal stresses around the rebars. Cracks will be formed in the concrete cover along the rebars, and finally concrete will be pushed off. In many practical cases, damage due to carbonation-induced reinforcement corrosion occurs faster than anticipated due to insufficient cover depths.

5.4.3 Chloride-induced corrosion

As mentioned in Chapter 1, mixed-in chlorides in too large quantities can cause reinforcement corrosion. However, more relevant for concrete practice are chlorides penetrating the concrete element from the environment, e.g. chlorides contained in sea water, industrial wastewaters or de-icing salts. The transport of chloride ions in concrete is a rather complex phenomenon, involving different mechanisms including capillary suction, diffusion, electrical migration, pressure-induced flow, wick action, and thermal migration (Yuan 2009, Audenaert et al. 2010).

Figure 5.45 Carbonation-induced corrosion in a column.

Penetration of chloride ions by capillary suction proceeds much faster than by diffusion. Capillary suction typically occurs in an outer layer of about 20 mm. This means that due to capillary suction, water containing chloride ions can very quickly reach a penetration depth of about 20 mm. From this capillary boundary on, chloride ions will typically further penetrate the concrete by diffusion. From a practical point of view, the diffusion process of chloride ions is governed by the same equations as for carbon dioxide. Fick's second law can be applied, showing that the chloride diffusion process, in the case of a constant diffusion coefficient, also follows a square root time relation.

Similar to the diffusion of carbon dioxide, in the case of chloride ions additional processes will also influence the diffusion process. Chlorides will be bound in hydration products, e.g. forming Friedel's salts $3CaO \cdot Al_2O_3 \cdot CaCl_2 \cdot 10H_2O$ in concrete based on C_3A-rich cement. While the aluminate phases play the most important role in *chloride binding* in concrete, other phases can also contribute. Binding of chloride in cementitious materials is a very complex process which is affected by many factors, including chloride concentration, cement composition, hydroxyl concentration, cation of chloride salt, temperature, and electrical field (Yuan et al. 2009). Previous carbonation of the concrete is also important in this respect, reducing the chloride-binding capacity.

It is also important to notice that following Fick's second law, diffusing chloride ions are treated as electrically neutral particles travelling in pore solution without the influence of other species. Nevertheless, chloride ions have a negative charge and, thus, a movement of chloride ions will be accompanied by a movement of some positive ions in the opposite direction. As a consequence, the cation type of chloride salt influences the chloride diffusion coefficient. Furthermore, the many types of ions present in the pore solution (Na^+, K^+, SO_4^{2-}, OH^-, ...) will also influence the diffusion process of chloride ions. In a more accurate chloride-transport model, the chloride ions should be treated as negative particles and the interaction with other ions has to be taken into account. For this kind of more advanced chloride-transport models, reference is made to literature (Yuan 2009).

Although the practically applied square root time relation is based on a constant diffusion coefficient, in reality, a time dependency exists and is often expressed in the following way with $D(t)$ the diffusion coefficient at time t, and D_{ref} the diffusion coefficient at reference time t_{ref} (Audenaert et al. 2010):

$$D(t) = D_{ref} \left(\frac{t_{ref}}{t} \right)^m \tag{5.25}$$

The exponent m is called the *age factor*, and is dependent on the concrete composition. According to Audenaert et al. (2010), the age factor m can be relatively well predicted as a function of the capillary porosity of the concrete with values of 0.4 for a capillary porosity equal to 5%, decreasing to 0.2 for a capillary porosity equal to 11%. Presently, the most widely applied diffusion models for chloride penetration into concrete consider a time dependent diffusion coefficient. Doing so, the solution to Fick's second law can be obtained as follows (Audenaert et al. 2010):

$$\frac{c}{c_0} = 1 - erf \left(\frac{x}{2 \sqrt{\dfrac{D_{ref}}{1-m} \left[\left(1 + \dfrac{t_{ex}}{\Delta t} \right)^{1-m} - \left(\dfrac{t_{ex}}{\Delta t} \right)^{1-m} \right] \left(\dfrac{t_{ref}}{t} \right)^m}} \right) \tag{5.26}$$

In this solution, c is the chloride concentration at a distance x from the exposed concrete surface; c_0 is the surface chloride concentration; t_{ex} is the age of the concrete at the start of exposure to chlorides; and Δt is the exposure duration. D_{ref} and t_{ref} are a known pair of diffusion coefficient and the age of the concrete, and m is the age factor as explained before.

With this prediction formula, based on chloride diffusion in an ageing concrete, the chloride penetration in real structures can be estimated in an acceptable way.

When a sufficient amount of chlorides reaches the steel reinforcement, the passivation layer can be broken down. This sufficient amount of chlorides is referred to as the *critical chloride content* or *chloride threshold concentration*. Different options are available to define the critical chloride content:

- based on the total chloride or on the water soluble ('free') chloride content, the difference between both being related to the process of chloride binding
- expressed as a threshold value relative to the mass of the concrete, or relative to the mass of cement within the concrete
- expressed as a critical [Cl-]/[OH-] ratio for the pore solution

The latter option seems to be the most fundamental because it also considers the alkalinity of the pore solution. A commonly considered critical [Cl-]/[OH-] ratio is 0.6. Nevertheless, for practical applications, this option is not very realistic due to experimental difficulties in determining the chloride concentration in the pore solution, which requires a complicated expression of pore solution. From a practical point of view, expressing the critical chloride content as a threshold value for the total chloride content seems to be the most preferable. Most typically, the threshold values are given relative to the mass of cement within the concrete. Unfortunately, no agreement exists on the threshold value to be considered, as reported values for the total chloride content relative to the mass of cement range between 0.17% and 2.5% (Audenaert 2006, Van den Bergh 2009). Many influencing parameters are said to be responsible for this wide variation, including type and amount of cement, chloride-binding capacity, porosity, temperature, type of cation accompanying the chlorides, oxygen content in the pore solution, chemical composition of the steel, and surface roughness of the steel. It can be said, however, that an often cited value is 0.4%.

The fact that no agreement has been reached on the chloride threshold concentration most probably is linked to the fact that the mechanism of the breakdown of the passive film layer by chloride ions is not yet fully understood (Rosenberg et al. 1989). Nevertheless, it seems generally accepted that the chloride ions become incorporated in the passive film, improving its solubility. Concerning the detailed mechanism, several theories have been formulated (Van den Bergh 2009). The details of these theories however go beyond the scope of this textbook.

It is important to stress, however, that the destruction of the passive layer typically occurs locally, rather than having an overall depassivation.

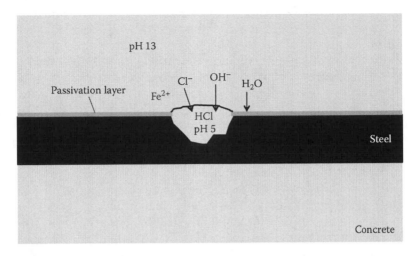

Figure 5.46 Chloride-induced pit corrosion of steel reinforcement in concrete.

As a result, the corrosion process of the steel can be very local, creating 'pit corrosion'. Within a pit, a microclimate can be obtained, as shown in Figure 5.46. The pH value in the pit decreases and more chloride ions are attracted, making the micro-environment even more corrosive. Iron ions Fe^{2+} react with chloride ions Cl^- and can be transported away from the reinforcement. After further oxidation to ferri-oxides and ferri-hydroxides, these corrosion products can become visible on the concrete surface, showing brownish colours. Furthermore, the expansive formation of corrosion products can also lead to cracking and spalling of the concrete cover, as is the case in carbonation-induced carbonation. However, due to pit corrosion, it is possible that the steel reinforcement fails due to an insufficient remaining cross section. This is a special risk in case of prestressing strands, which are more sensitive to chloride-induced corrosion than classical reinforcement. Prestressing strands or wires can fail due to pit corrosion, even before corrosion products become visible on the concrete surface. Due attention should be given to this risk when applying excessive amounts of de-icing salts on prestressed road bridges.

In this context, *stress-corrosion cracking and hydrogen embrittlement* can also be mentioned, which can be considered a synergistic action between mechanical and environmental loading. Prestressing steel is particularly susceptible to this kind of attack, although failure typically also requires insufficient concrete quality, inadequate mortar injection of the cable ducts, or corrosion prior to injection (Rosenberg 1989). In cathodic regions, hydrogen is formed, which can penetrate the steel. As a result, the steel lattice structure is distorted and the steel becomes more brittle. Furthermore, atomic hydrogen (H) can be transformed to molecular hydrogen (H_2), causing high internal pressures in the steel and possibly crack initiation. The role of chlorides in this process is twofold, as pit corrosion on the one hand leads to hydrogen formation and, on the other hand, leads to stress concentrations in the mechanically loaded steel.

In order to *prevent chloride-induced corrosion* of steel reinforcement, the same basic measures have to be taken as for carbonation-induced corrosion: a good quality concrete has to protect the reinforcement from the aggressive environment, and an adequate concrete cover thickness must be considered. Appropriate curing of the freshly cast concrete is important in order to reach the required concrete quality in the cover zone. However, when considering the effect of binder type, a different situation is to be noticed in comparison with carbonation. As the chloride-binding capacity is higher in the case of blast furnace slag cement, this cement type might be preferred in case of chloride attack instead of Portland cement; whereas in the case of carbonation-induced corrosion, a Portland cement will typically perform better. However, a good curing of the freshly cast concrete is stressed again because lack of curing will increase the carbonation rate in the case of blast furnace slag cement, which will further increase the chloride penetration.

Chloride-induced corrosion of reinforcing steel in concrete can often be found in marine structures. An example is shown in Figure 5.47 for the case of some concrete elements near the Adriatic Sea. Severe damage is

Figure 5.47 Chloride induced reinforcement corrosion in some concrete elements near the Adriatic Sea.

already visible, with large areas of concrete cover pushed off by the expansive corrosion products. Again, it can be seen that the extensive corrosion in this case is most probably related to insufficient cover thickness.

REFERENCES

ACI Committee 210 (2003) 'Erosion of concrete in hydraulic structures', ACI 210R-03, American Concrete Institute, pp.24.

ACI Committee 435 (1997) 'State-of-the-art report on temperature-induced deflections of reinforced concrete members', ACI 435.7R-85, pp.14.

Annerel E. (2010) 'Assessment of the residual strength of concrete structures after fire exposure', Doctoral thesis, Magnel Laboratory for Concrete Research, Ghent University, Belgium, pp.291.

Aroni S. (1993) 'Autoclaved aerated concrete: properties, testing, and design', RILEM Recommended Practice, Spon Press, London an New York, ISBN 0-419-17960-7.

Asgeirsson H. (ed.) (1975) 'Proceedings of a symposium on alkali-aggregate reaction, preventive measures', Icelandic Building Research Institute, Reykjavik, Iceland.

Audenaert K. (2006) 'Transport mechanisms in self-compacting concrete in relation with carbonation and chloride penetration' (in Dutch), Doctoral thesis, Magnel Laboratory for Concrete Research, Ghent University, Belgium, pp.369.

Audenaert K., Yuan Q. and De Schutter G. (2010) 'On the time dependency of the chloride migration coefficient in concrete', Construction and Building Materials, 24, 396–402.

Bassuoni M.T. and Nehdi M.L. (2009) 'Durability of self-consolidating concrete to sulfate attack under combined cyclic environments and flexural loading', Cement and Concrete Research, 39, 206–226.

Bazant Z.P. and Steffens A. (2000) 'Mathematical model for kinetics of alkali-silica reaction in concrete', Cement and Concrete Research, 30, 419–428.

Beaudoin J.J. and McInnis C. (1974) 'The mechanism of frost damage in hardened cement paste', Cement and Concrete Research, 4, 2, 139–147.

Bellmann F., Erfurt A. and Ludwig H.-M. (2012) 'Field performance of concrete exposed to sulphate and low pH conditions from natural and industrial sources', Cement and Concrete Composites, 34, 86–93.

Boel V., Cnudde V., De Schutter G., Van Meel B., Masschaele B., Ye G., Van Hoorebeke L., Jacobs P. (2008) 'X-ray computed microtomography on cementitious materials: possibilities and limitations', in 'Concrete Modelling CONMOD'08', Eds. E. Schlangen and G. De Schutter, RILEM Proceedings PRO 58, RILEM Publications S.A.R.L., ISBN 978-2-35158-060-8, 487–494.

Boström L. and Jansson R. (2007) 'Fire resistance', in Durability of Self-Compacting Concrete, eds. De Schutter G. and Audenaert K., RILEM State-of-the-art Report 38, RILEM Publications S.A.R.L., ISBN 978-2-35158-048-6, 143–152.

Browne R.D. and Bamforth P.B. (1982) 'The use of concrete for cryogenic storage: a summary of research, past and present', in 'Cryogenic Concrete', Proceedings of the 1st International Conference, Newcastle upon Tyne, March 1981, Construction Press, London and New York, ISBN 0-86095-705-5, 135–162.

Buettner D.R. and Becker R.J. (1998) *PCI Manual for the design of hollow core slabs*, 2nd edition, Precast Concrete Institute, Chicago, Illinois.

Bulteel D., Garcia-Diaz E. and Dégrugilliers P. (2010) 'Influence of lithium hydroxide on alkali–silica reaction', *Cement and Concrete Research*, 40, 526–530.

Chatterji S. (2003) 'Freezing of air-entrained cement-based materials and specific actions of air-entraining agents', *Cement and Concrete Research*, 25, 759–765.

Clifton J.R. and Ponnersheim J.M. (1994) 'Sulfate attack of cementitious materials: volumetric relations and expansions', NISTIR 5390, NIST, Gaithersburg, pp.18.

De Belie N. (1997) 'Concrete technological and chemical aspects of floor degradation in pig houses' (in Dutch), doctoral thesis, Magnel Laboratory for Concrete Research, Ghent University, Belgium.

De Ceukelaire L. (1986) Report 86/0704 of the Magnel Laboratory for Concrete Research, Ghent University, Belgium.

De Ceukelaire L. (1988) 'Alkali-silicareactie nu ook in België?', *Cement*, 10, 21–25.

De Ceukelaire L. (1991) 'The determination of the most common crystalline alkali-silica reaction product', *Materials and Structures*, 24, 169–171.

Fares H., Rémond S., Noumowé A. and De Schutter G. (2011) 'Comportement à haute température des bétons auto-plaçants', in 'Les bétons auto-plaçants', Ed. A. Loukili, Lavoisier, Paris, ISBN 978-2-7462-3127-6, 2011, 219–260.

Feng X, Thomas M.D.A., Bremner T.W., Folliard K.J. and Fournier B. (2010) 'Summary of research on the effect of $LiNO_3$ on alkali–silica reaction in new concrete', *Cement and Concrete Research*, 40, 636–642.

Ferraris C.F. (1995) 'Alkali-silica reaction and high performance concrete', National Institute of Standards and Technology, NISTIR 5742, pp.20.

Grattan-Bellew P.E., Mitchell L.D., Margeson J. and Min D. (2010), 'Is alkali-carbonate reaction just a variant of alkali silica reaction ACR=ASR ?', *Cement and Concrete Research*, 40, 556–562.

Gruyaert E., Van den Heede Ph., Maes M. and De Belie N. (2012) 'Investigation of the influence of blast-furnace slag on the resistance of concrete against organic acid or sulphate attack by means of accelerated degradation tests', *Cement and Concrete Research*, 42, 173–185.

Haynes H. and Bassuoni M.T. (2011) 'Physical salt attack on concrete', *Concrete International*, 33, 11, 38–42.

Helmuth R., Stark D., Diamond S. and Moranville-Regourd M. (1993) 'Alkali-silica reactivity: an overview of research', Strategic Highway Research Program, SHRP-C-342, National Research Council, Washington DC, ISBN 0-309-05602-0.

Hime W.G. (2003) 'Chemists should be studying chemical attack on concrete', *Concrete International*, 25, 4, 82–84.

Hobbs D.W. (1988) 'Alkali-silica reaction in concrete', Thomas Telford, London.

Ichikawa T. and Miura M. (2007) 'Modified model of alkali-silica reaction', *Cement and Concrete Research*, 37, 1291–1297.

Irassar E.F. (2009) 'Sulfate attack on cementitious materials containing limestone filler – a review', *Cement and Concrete Research*, 39, 241–254.

Katayama T. (1992) 'A critical review of carbonate rock reactions – is their reactivity useful or harmful ?', Proceedings of the 9th International Conference on alkali-aggregate reaction in concrete (ICAAR), ed. A.B. Poole, London, 508–517.

Katayama T. (2010) 'The so-called alkali-carbonate reaction (ACR) – Its mineralogical and geochemical details, with special reference to ASR', *Cement and Concrete Research*, 40, 643–675.

Khoury G.A. (2000) 'Effects of fire on concrete and concrete structures', *Prog. Struct. Eng. Mat.*, 429–447.

Kohler S., Heinz D. and Urbonas L. (2006) 'Effect of ettringite on thaumasite formation', *Cement and Concrete Research*, 36, 697–706.

Larreur-Cayol S., Bertron A. and Escadeillas G. (2011) 'Degradation of cement-based materials by various organic acids in agro-industrial waste-waters', *Cement and Concrete Research*, 41, 882–892.

Leemann A., Lothenbach B. and Hoffmann C. (2010) 'Biologically induced concrete deterioration in a wastewater treatment plant assessed by combining microstructural analysis with thermodynamic modeling', *Cement and Concrete Research*, 40, 1157–1164.

Liu X. (2006) 'Microstructural Investigation of Self-Compacting Concrete and High-Performance Concrete during Hydration and after Exposure to High Temperatures', Doctoral thesis, Magnel Laboratory for Concrete Research, Ghent University, Belgium & Tongji University, Shanghai, PR China, pp.174.

Liu X., Ye G., De Schutter G., Yuan Y. and Taerwe L. (2008) 'On the mechanism of polypropylene fibres in preventing fire spalling in self-compacting and high-performance cement paste', *Cement and Concrete Research*, 38, 487–499.

Liu Y.-W., Yen T. and Hsu T.-H. (2006) 'Abrasion erosion of concrete by water-borne sand', *Cement and Concrete Research*, 36, 1814–1820.

Liu Z. (2010) 'Study of the basic mechanisms of sulfate attack on cementitious materials', Doctoral thesis, Magnel Laboratory for Concrete Research, Ghent University, Belgium & Central South University, Changsha, PR China, pp.190.

Liu Z., De Schutter G., Deng D. and Yu Z. (2010) 'Micro-analysis of the role of interfacial transition zone in "salt weathering" on concrete', *Construction and Building Materials*, 24, 2052–2059.

Liu Z., Deng D., De Schutter G. and Yu Z. (2011) 'Micro-analysis of "salt weathering" on cement paste', *Cement and Concrete Composites*, 33, 179–191.

Liu Z., Deng D., De Schutter G. and Yu Z. (2012) 'Chemical sulfate attack performance of partially exposed cement and cement + fly ash paste', *Construction and Building Materials*, 28, 230–237.

Lottman B.B.G., Koenders E.A.B and Walraven J.C. (2011) 'Macro-scale spalling model: a fracture mechanics versus pore pressure approach', in Concrete Spalling due to Fire Exposure, Koenders and Dehn (eds.), RILEM Proceedings PRO 80, RILEM Publications S.A.R.L., ISBN 978-2-35158-118-6, 53–65.

Marchand J., Samson E., Maltais Y. and Beaudoin J.J. (2002) 'Theoretical analysis of the effect of weak sodium sulfate solutions on the durability of concrete', *Cement and Concrete Composites*, 24, 317–329.

Matsushita F., Aono Y. and Shibata S. (2004a) 'Microstructure changes in autoclaved aerated concrete during carbonation under working and accelerated conditions', *Journal of Advanced Concrete Technology*, Vol. 2, No.1, 121–129.

Matsushita F., Aono Y. and Shibata S. (2004b) 'Calcium silicate structure and carbonation shrinkage of a tobermorite-based material', *Cement and Concrete Research*, 34, 1251–1257.

Matsushita F., Aono Y. and Shibata S. (2009) 'Carbonation shrinkage mechanism of a tobermorite-based material', in *Creep, Shrinkage and Durability Mechanisms of Concrete and Concrete Structures*, Tanabe et al. (Eds.), Taylor and Francis Group, London, ISBN 978-0-415-48508-1, 41–47.

Mehta P.K. (2000) 'Sulfate attack on concrete: separating myths from reality', *Concrete International*, 22, 8, 57–61.

Monteny J., Vincke E., Beeldens A., De Belie N., Taerwe L., Van Gemert D. and Verstraete W. (2000) 'Chemical, microbiological, and in situ test methods for biogenic sulphuric acid corrosion on concrete', *Cement and Concrete Research*, 30, 4, 623–634.

Moskvin V.M. (1978) 'Cavitation erosion of concrete', *Power Technology and Engineering*, Vol. 12, No.12, 1215–1218.

Multon S., Cyr M., Sellier A., Diederich P. and Petit L. (2010) 'Effects of aggregate size and alkali content on ASR expansion', *Cement and Concrete Research*, 40, 4, 508–516.

Neville A. (2004) 'The confused world of sulfate attack on concrete', *Cement and Concrete Research*, 34, 1275–1296.

Neville A.M. and Brooks J.J. (2010) *Concrete Technology*, Second Edition, Prentice Hall, Pearson, ISBN 978-0-273-73219-8.

Mehta P.K. and Monteiro P.J.M. (2006) *Concrete: Microstructure, properties and materials*, McGraw-Hill, Third Edition, ISBN 0-07-146289-9.

O'Connell M., McNally C. and Richardson M.G. (2010) 'Biochemical attack on concrete in wastewater applications: A state of the art review', *Cement and Concrete Composites*, 32, 479–485.

Papenfus N. (2003) 'Applying concrete technology to abrasion resistance', Proceedings of the 7th International Conference on Block Paving, South Africa, ISBN 0-958-46091-4, pp.9.

Pel L., Huinink H., Kopinga K. and Van Hees R.P.J. (2004) 'Efflorescence pathway diagram: understanding salt weathering', *Construction and Building Materials*, 18, 5, 309–313.

Phan L.T. and Carino N.J. (2000) 'Fire performance of high strength concrete: research needs', in Advanced Technology in Structural Engineerg, ASCE/SEI Structures Congress, Ed. Algaaly E., pp.8.

Pourbaix M. (1976) 'Atlas of electrochemical equilibrium in aqueous solutions', Pergamon, London.

Powers T.C. (1945) 'A working hypothesis for further studies of frost resistance of concrete', *Journal of the American Concrete Institute*, 16, 4, 245–272.

Powers T.C. and Steinour H.H. (1955) 'An interpretation of some published reachearchers on alkali-aggregate reaction. Part 1 – The chemical reactions and mechanisms of expansion', *Journal of the American Concrete Institute*, 26, 497–516.

Powers T.C. and Steinour H.H. (1955) 'An interpretation of some published reachearchers on alkali-aggregate reaction. Part 2 – A hypothesis concerning safe and unsafe reactions with reactive silica in concrete', *Journal of the American Concrete Institute*, 26, 785–811.

Powers T.C. (1975) 'Freezing effects in concrete', in Durability of Concrete, ACI SP-47, 1–11.

Rosenberg A., Hansson C.M. and Andrade C. (1989) 'Mechanisms of corrosion of steel in concrete', in *Materials Science of Concrete I* (ed. J. Skalny), The American Ceramic Society, ISBN 0-944904-01-7, 285–313.

Rostasy F.S., Schneider U. and Wiederman G. (1979) 'Behaviour of mortar and concrete at extremely low temperatures', *Cement and Concrete Research*, Volume 9, Issue 3, 365–376.

Rostasy F.S. and Wiederman G. (1980) 'Stress-strain behaviour of concrete at extremely low temperature', *Cement and Concrete Research*, Volume 10, Issue 4, 565–572.

Schiessl P. (1976) 'Zur Frage der zulassigen Rissbreite und der erforderlichen Betondeckung im Stahlbetonbau unter besonderer Berucksichtigung der Karbonatisierung des Betons', Deutscher Ausschuss für Stahlbeton, Berlin, Heft 255.

Schmidt T., Lothenbach B., Romer M., Neuenschwander J. and Scrivener K. (2009) 'Physical and microstructural aspects of sulfate attack on ordinary and limestone blended Portland cements', *Cement and Concrete Research*, 39, 1111–1121.

Setzer M. (2003) 'Micro-ice-lens formation in porous solid', *Journal of Colloid and Interface Science*, 243, 1, 193–201.

Setzer M., Janssen D.J., Auberg R. et al (2004) 'Final report of RILEM TC 176-IDC: Internal damage of concrete due to frost action', *Materials and Structures*, 37, 274, 740–742.

Shehata M.H. and Thomas M.D.A. (2000) 'The effect of fly ash composition on the expansion of concrete due to alkali silica reaction', *Cement and Concrete Research*, 30, 1063–1072.

Shehata M.H. and Thomas M.D.A. (2002) 'Use of ternary blends containing silica fume and fly ash to suppress expansion due to alkali-silica reaction in concrete', *Cement and Concrete Research*, 32, 341–349.

Skalny J., Marchand J. and Odler I. (2002) *Sulfate attack on concrete*, Sponn Press, London, ISBN 0-419-24550-2.

Sleigh G. (1981) 'The behaviour of steel at low temperatures', in 'Cryogenic Concrete', Proceedings of the 1st International Conference, Newcastle upon Tyne, March 1981, Construction Press, London and New York, ISBN 0-86095-705-5, 167–177.

Soutsos M. (ed.) (2010) *Concrete durability. A practical guide to the design of durable concrete structures*, Thomas Telford, London, ISBN 978-0-7277-3517-1.

Stark J. (2002) 'Performance of concrete in sulfate environments', Research and Development Bulletin, RD 097, Portland Cement Association, Skokie.

Sun Z. and Scherer G.W. (2010) 'Effect of air voids on salt scaling and internal freezing', *Cement and Concrete Research*, 40, 260–270.

Swamy R.N. (ed.) (1992) *The alkali-silica reaction in concrete*, Blackie and Son Ltd., Glasgow and London, ISBN 0-216-92691-2.

The Concrete Society (1981) 'Cryogenic Concrete', Proceedings of the 1st International Conference, Newcastle upon Tyne, March 1981, Construction Press, London and New York, ISBN 0-86095-705-5.

Thomas M. (2011) 'The effect of supplementary cementing materials on alkali-silica reaction: a review', *Cement and Concrete Research*, 41, 1224–1231.

Tuutti K. (1982) 'Corrosion of steel in concrete', Swedish Cement and Concrete Research Institute, Stockholm.

Ueda T., Baba Y. and Nanasawa A. (2011) 'Effect of electrochemical penetration of lithium ions on concrete expansion due to ASR', *Journal of Advanced Concrete Technology*, 9, 1, 31–39.

Valenza J.J. (2005) 'Mechanism for salt scaling', PhD thesis, Princeton, NJ.

Valenza J.J. and Scherer G.W. (2007a) 'A review of salt scaling: I. Phenomenology', *Cement and Concrete Research*, 37, 1007–1021.

Valenza J.J. and Scherer G.W. (2007b) 'A review of salt scaling: II. Mechanisms', *Cement and Concrete Research*, 37, 1022–1034.

Van den Bergh K. (2009) 'Study of the corrosion of reinforcing steel in concrete by means of electrochemical methods' (in Dutch), Doctoral thesis, Free University Brussels, pp.267.

van der Veen C. (1987) 'Properties of concrete at very low temperatures; A survey of the literature', Stevin Laboratory, Faculty of Civil Engineering, Delft University of Technology, Report N° 25-87-2.

Waagaard K. (1981) 'Developments in marine applications', in 'Cryogenic Concrete', Proceedings of the 1st International Conference, Newcastle upon Tyne, March 1981, Construction Press, London and New York, ISBN 0-86095-705-5, 108–121.

Wiedemann G. (1982) 'Zum einfluss tiefer Temperaturen auf Festigkeit und Verformung von Beton', Dissertation Technische Universität Braunschweig, pp.149.

Yamane S., Kasami H. and Okuno T. (1978) 'Properties of concrete at very low temperatures', ACI Special Publication 55, 207–222.

Ye G., Liu X., De Schutter G., Taerwe L. and Vandevelde P. (2007) 'Phase distribution and microstructural changes of self-compacting cement paste at elevated temperature', *Cement and Concrete Research*, 37, 978–987.

Yuan Q. (2009) 'Fundamental studies on test methods for the transport of chloride ions in cementitious materials', Doctoral thesis, Magnel Laboratory for Concrete Research, Ghent University, Belgium & Central South University, Changsha, PR China, pp.340.

Yuan Q., Shi C., De Schutter G., Audenaert K. and Deng D. (2009) 'Chloride binding of cement-based materials subjected to external chloride environment – A review', *Construction and Building Materials*, 23, 1–13.

Index